英語で学ぶ
数 値 解 析
Numerical Analysis

陳　　小君
山本　哲朗　共著

コロナ社

まえがき

　現代は国際化の時代であって，好むと好まざるとにかかわらず，至るところで英語の必要性が増しています．大学も例外ではなく，近年，外国の多くの大学で，専門科目の講義をすべて英語で行う計画が進行中です．この傾向はやがてわが国にも及ぶでしょう．本書はこのような時代潮流を見据え，理工系学生にとって必要な数値解析の基礎を英文で記述したものです．これにより数学的素養を身につけ，かつ英文読解力の養成を図るという一石二鳥をねらうとともに，理工系専門英語のテキストとしても使えるよう意図しています．

　その他本書の特徴はつぎのとおりです．

1. 数値解析の基礎事項を精選し，必須事項のみを取り上げました．したがって，高度な内容は含めていません．しかし，各事項の理解を容易にするために叙述に工夫を凝らしました．すべてのアルゴリズムに例題と計算例をつけたのもこのためです．さらに，各章末に演習問題をつけ，解答を和文で末尾に記しました．
2. 英文はなるべくやさしい構文を用い，各ページに脚注として専門用語の訳を記しました．また，読者の便宜を考え，巻末に英和と和英の両索引を付しました．なお，英文の校閲を，著者たちの長年の友人である，元 Wisconsin 大学数学科教授 Louis B. Rall 氏にお願いし，万全を期しました．ただし，内容については著者たちが全責任を負うものです．
3. 本書のレベルはわが国の大学の理工系学部学生程度を意図していますが，数値解析の標準的教科書として，国際的にも十分通用するものと自負しています．

　本書により読者が英文教科書に慣れ親しみ，かつ，数値解析に興味を抱いていただければ，著者たちの喜びこれに優るものはありません．

　最後に，原稿を通読され，完全な英文の作成にご協力いただいた Louis B. Rall 氏と執筆中有益な助言をいただいた僚友土屋卓也，方青両氏および本書の出版をご快諾いただいたコロナ社に対し深甚なる謝意を表します．

2002 年 8 月

　　　　　　　　　　　　　　　　　　　　　　　　陳　小君・山本哲朗

本書を教科書として使用される方へ

1. 本書は1年間の講義に相当する内容を含みますが，各章が独立しているため，講義を担当される方々の計画に応じて，半年間の講義または集中講義のために一部分をとり出して使うことができます．ただし，6章と7章を理解するためには5章の知識を必要とします．
2. 数学的厳密さを失わぬよう，本文に若干の収束定理と誤差評価の証明を付していますが，場合によっては，アルゴリズムと例題を中心として，証明を省略して使うことができます．アルゴリズムの理解が十分できた後に数学的理論を学んでもよいのではないかと著者達は考えるからです．

目次

まえがき
目次

1 数値計算における誤差
- 1.1 絶対誤差と相対誤差 ... 1
- 1.2 丸めの誤差 ... 2
- 1.3 打切り誤差 ... 4
- 1.4 誤差の伝播 ... 6
- 1.5 有効けた ... 7
- 演習問題 ... 9

2 連立線形方程式
- 2.1 序 ... 10
- 2.2 ガウス消去法 ... 11
- 2.3 逆行列の計算 ... 16
- 2.4 LU 分解法 ... 17
- 2.5 反復法 ... 20
- 2.6 ベクトルノルムと行列ノルム ... 23
- 2.7 反復法の収束性 ... 29
- 演習問題 ... 34

3 非線形方程式
- 3.1 序 ... 36
- 3.2 不動点法 ... 37
- 3.3 ニュートン法 ... 41
- 3.4 セカント法 ... 43
- 3.5 連立非線形方程式 ... 45
- 3.6 多項式の零点 ... 49
- 演習問題 ... 52

4 行列の固有値問題
- 4.1 固有値と固有ベクトル ... 53
- 4.2 固有値の包含定理 ... 55
- 4.3 累乗法 ... 57

目次

- 4.4 ハウスホルダー行列と3重対角化 ... 59
- 4.5 QR分解法 ... 63
- 演習問題 ... 65

5 補間多項式
- 5.1 序 ... 67
- 5.2 ラグランジュの補間公式 ... 68
- 5.3 ニュートンの補間公式 ... 71
- 5.4 ニュートンの前進および後退補間公式 ... 75
- 5.5 差分商の拡張 ... 78
- 5.6 誤差公式 ... 83
- 演習問題 ... 85

6 数値積分
- 6.1 序 ... 86
- 6.2 中点公式 ... 89
- 6.3 台形公式 ... 92
- 6.4 シンプソン公式 ... 95
- 演習問題 ... 99

7 常微分方程式の初期値問題
- 7.1 序 ... 101
- 7.2 オイラー法 ... 102
- 7.3 ルンゲ・クッタ法 ... 105
- 7.4 アダムス・バッシュフォース法 ... 109
- 演習問題 ... 113

8 微分方程式に対する差分法
- 8.1 2点境界値問題 ... 115
- 8.2 楕円型方程式 ... 121
- 8.3 放物型方程式 ... 127
- 8.4 双曲型方程式 ... 131
- 演習問題 ... 136

付録A 数学式の読み方 ... 138
付録B ギリシャ文字 ... 142
参考文献 ... 143
演習問題略解 ... 144
英和索引 ... 148
和英索引 ... 153

Preface

This book provides a fundamental introduction to numerical analysis suitable for undergraduate students in mathematics, physics, computer science, and engineering. In particular, the English used in this book is very simple so that the reader whose second language is English can read it without language problems. In fact, this book has been used at several universities in Japan for either a single-term course or a year course.

In order to make the book accessible to scientists and engineers, we have written it to be as self-contained as possible. It is only assumed that the reader is familiar with calculus and linear algebra. Error analysis and convergence analysis for each numerical method are given by using elementary mathematics. Moreover, there is a wide variety of examples that help the reader to understand the numerical methods and the theory in numerical analysis. Moreover, all examples can be carried out by hand so that they can be used in a class room.

The mathematical content of the book is comparable with standard textbooks of numerical analysis for undergraduate courses in universities of America, Australia and Germany.

The text and the more elementary figures of the book were typeset by the authors using LaTeX. The more complicated figures were created using Matlab.

We would like to thank Professor Louis B. Rall, University of Wisconsin, for carefully reading this book and for correcting use of the English language. Any remaining defects are the responsibility of the authors.

Finally we thank our colleagues Takuya Tsuchiya and Qing Fang of Ehime University for their helpful suggestions.

Xiaojun Chen
Tetsuro Yamamoto

August, 2002

Contents

Chapter 1 Errors in Numerical Computation
- 1.1 Absolute errors and relative errors 1
- 1.2 Round-off errors 2
- 1.3 Truncation errors 4
- 1.4 Error propagation 6
- 1.5 Significant digits 7
- Exercises 9

Chapter 2 Systems of Linear Equations
- 2.1 Introduction 10
- 2.2 Gaussian elimination 11
- 2.3 Matrix inversion 16
- 2.4 LU-factorization 17
- 2.5 Iterative methods 20
- 2.6 Vector norms and matrix norms 23
- 2.7 Convergence of iterative methods 29
- Exercises 34

Chapter 3 Systems of Nonlinear Equations
- 3.1 Introduction 36
- 3.2 The fixed point method 37
- 3.3 Newton's method 41
- 3.4 The secant method 43
- 3.5 Systems of nonlinear equations 45
- 3.6 Zeros of polynomials 49
- Exercises 52

Chapter 4 The Matrix Eigenvalue Problem
- 4.1 Eigenvalues and eigenvectors 53
- 4.2 Inclusion of eigenvalues 55
- 4.3 The power method 57
- 4.4 Householder tridiagonalization 59
- 4.5 The QR-factorization method 63
- Exercises 65

Chapter 5   Interpolation Polynomials
- 5.1 Introduction 67
- 5.2 The Lagrange interpolation formula 68
- 5.3 The Newton divided difference interpolation formula 71
- 5.4 The Newton forward and backward interpolation formulas . 75
- 5.5 Extension of divided differences 78
- 5.6 Error formula 83
- Exercises 85

Chapter 6   Numerical Integration
- 6.1 Introduction 86
- 6.2 The rectangular rule 89
- 6.3 The trapezoidal rule 92
- 6.4 Simpson's rule 95
- Exercises 99

Chapter 7   Initial Value Problems for Ordinary Differential Equations
- 7.1 Introduction 101
- 7.2 Euler's method 102
- 7.3 Runge-Kutta methods 105
- 7.4 Adams-Bashforth methods 109
- Exercises 113

Chapter 8   Finite Difference Methods for Differential Equations
- 8.1 Two-point boundary value problems 115
- 8.2 Elliptic equations 121
- 8.3 Parabolic equations 127
- 8.4 Hyperbolic equations 131
- Exercises 136

Appendix A How to Read Mathematical Expressions in English 138
Appendix B Greek Alphabet 142
References 143
Answers to Selected Exercises 144
Index (English-Japanese) 148
Index (Japanese-English) 153

Chapter 1.

Errors in Numerical Computation

1.1 Absolute errors and relative errors

In the practice of numerical computation, it is important to be aware that computed results are not exact mathematical results. If $\hat{x}$ is an approximate value of a true value x, then the error of $\hat{x}$ is

$$e = x - \hat{x}, \qquad \text{Error= True value} - \text{Approximate value} .$$

The **absolute error** of $\hat{x}$ is defined by

$$|e| = |x - \hat{x}|, \qquad |\text{Error}|.$$

The **relative error** e_R of $\hat{x}$ is defined by

$$e_R = \frac{|e|}{x} = \frac{|x - \hat{x}|}{x}, \qquad \frac{|\text{Error}|}{\text{True value}}.$$

In practice, the absolute error and relative error can not be computed, because the true value x is unknown. What we can compute is a number ϵ or ϵ_R such that

$$|e| \leq \epsilon, \quad \text{or} \quad |e_R| \leq \epsilon_R.$$

---

absolute error [絶対誤差]　　relative error [相対誤差]　　true value [真値]　　approximate value [近似値]

Such numbers ϵ and ϵ_R are called **error bounds** for $\hat{x}$. The error bound for the absolute error $|e|$ gives a computable interval which includes the true value x, that is,
$$x \in [\hat{x} - \epsilon, \ \hat{x} + \epsilon].$$
For the relative error, if ϵ is much less than $|\hat{x}|$, then we can use $\hat{x}$ instead of x in e_R and get
$$|e_R| = \frac{|e|}{|\hat{x} + e|} \leq \frac{\epsilon}{|\hat{x}| - \epsilon} = \frac{\epsilon}{|\hat{x}|} + \left(\frac{\epsilon}{|\hat{x}|}\right)^2 + \ldots \approx \frac{\epsilon}{|\hat{x}|}.$$

Example 1.1 Let $\hat{x} = 3.142$ be an approximate value of $x = 3.1416$. Find the absolute error and relative error of $\hat{x}$. Give an error bound for $\hat{x}$ if $\hat{x}$ is an approximate value of π.

Solution The absolute error of $\hat{x}$ is
$$|e| = |x - \hat{x}| = |3.1416 - 3.142| = 0.0004.$$
The relative error is
$$e_R = \frac{|x - \hat{x}|}{x} = \frac{0.0004}{3.1416}.$$
If $\hat{x}$ is an approximate value of π, we can not compute the absolute error and the relative error of $\hat{x}$, because the exact value of π is unknown. Now we show that $\epsilon = 0.0005$ is an error bound for the absolute error and $\epsilon_R = 0.0005/3.1415$ is an error bound for the relative error.

Since $\hat{x} > \pi > 3.1415$, we have
$$|\pi - \hat{x}| = \hat{x} - \pi \leq \hat{x} - 3.1415 = 0.0005$$
and
$$\frac{|\pi - \hat{x}|}{\pi} \leq \frac{0.0005}{\pi} \leq \frac{0.0005}{3.1415}. \qquad \diamond$$

1.2 Round-off errors

In numerical computation, we use numbers not with infinite precision but rather with a finite number of digits. A round-off error is caused by discarding all digits from some digit on.

Consider a number that has m digits
$$x = \pm 0.d_1 d_2 \ldots d_m.$$

---

error bound [誤差限界]　　round-off error [丸め誤差]　　digit [けた]

1.2 Round-off errors

The general rule for rounding off x to a number $\hat{x}$ with k digits is illustrated as follows. Let
$$b = \pm 0.0\ldots 0 d_{k+1}\ldots d_m$$
and
$$c = \frac{1}{2} \times 0.\overbrace{0\cdots 0}^{k-1}1.$$
The number b is to be discarded, and c is half a unit in the kth place.

(a) If $|b| < c$, let $\hat{x} = \pm 0.d_1 d_2 \ldots d_k$, that is, leave the kth digit unchanged ("rounding down").
(b) If $|b| > c$, let $\hat{x} = \pm (0.d_1 d_2 \ldots d_k + 2c)$, that is, add one to the kth digit ("rounding up").
(c) If $|b| = c$, round off to the nearest *even* digit. (Example: Rounding off 0.125 and 0.135 to 2 digits gives 0.12 and 0.14, respectively.)

The rule (c) is supposed to ensure that in discarding exactly half a digit, rounding up and rounding down happen about equally often, on the average.

If we round off 3.1415 to 4 digits, we get 3.142.
If we round off 3.1425 to 4 digits, we also get 3.142.
If we round off 3.4535 to 4,3,2 digits, we get 3.454, 3.45, 3.5.
If we round off 3.45 to 2 digit, we get 3.4.

Many computers have several special rules for rounding off a number to k digits.

- Round a number to the nearest k digits number towards minus infinity. (Example: Rounding off 3.1418 to 4 digits gives 3.141)
- Round a number to the nearest k digit number towards infinity. (Example: Rounding off 3.1412 to 4 digits gives 3.142)
- Round a number to the nearest k digit number towards zero. (Example: Rounding off 3.1412 to 4 digits gives 3.141)

We can specify one or more round-off rules before computation. It is notable that many numerical verification programs use more than one round-off rule to obtain intervals including true values. For example, rounding π to the nearest 4 digits number towards infinity and minus infinity respectively gives 3.142 and 3.141 which provides an interval $[3.141, 3.142]$ containing the exact value of π.

rounding down [切捨て]　rounding up [切上げ]　interval [区間]　average [平均]　infinity [無限]

1.3 Truncation errors

A truncation error is caused by replacing a complicated function with the sum of the first few terms of its Taylor series. For example, replacing the infinite Taylor series

$$f(x) := e^x = 1 + x + \frac{x^2}{2!} + \frac{x^3}{3!} + \ldots, \quad x > 0$$

by the sum of its first $n+1$ terms

$$f_n(x) = 1 + x + \frac{x^2}{2!} + \ldots + \frac{x^n}{n!},$$

causes the truncation error

$$e_T = f(x) - f_n(x) = \frac{e^\xi x^{n+1}}{(n+1)!}, \quad 0 < \xi < x.$$

In numerical methods for solving differential equations, a truncation error is caused when we replace the derivatives by the finite difference formulas. For example, replacing $f'(x)$ by $(f(x+h) - f(x))/h$ causes the trucation error

$$e_T = f'(x) - \frac{f(x+h) - f(x)}{h} = -\frac{1}{2} f''(\xi) h, \quad \xi = x + \theta h, \ 0 < \theta < 1.$$

This follows from the Taylor expansion

$$f(x+h) = f(x) + f'(x)h + \frac{1}{2} f''(\xi) h^2.$$

We say that a function $f(h)$ is approximated by the function $g(h)$ with order of approximation $O(h^n)$ if there exist a constant $C > 0$ and a positive integer n such that

$$\frac{|f(h) - g(h)|}{h^n} \leq C \quad \text{as } h \to 0.$$

In this case, we write

$$f(h) = g(h) + O(h^n).$$

If

$$\frac{|f(h) - g(h)|}{h^n} \to 0 \quad \text{as } h \to 0,$$

---

truncation error [打切り誤差]　　Taylor series [テイラー級数]　　derivative [導関数]
finite differnce [有限差分]　　Taylor expansion [テイラー展開]

1.3 Truncation errors

then we write
$$f(h) = g(h) + o(h^n).$$

Assume that $f(h) = g(h) + O(h^n)$, $p(h) = q(h) + O(h^m)$ and $k = \min(n, m)$. Then it is not difficult to verify the following results on combinations of two functions:

$$f(h) + p(h) = g(h) + q(h) + O(h^k)$$

$$f(h)p(h) = g(h)q(h) + O(h^k),$$

and
$$\frac{f(h)}{p(h)} = \frac{g(h)}{q(h)} + O(h^k), \quad \text{if } p(h) \neq 0, \ q(h) \neq 0,$$

provided that $f, g, p, q, p^{-1}, q^{-1}$ are bounded.

Using this notation, we can write
$$e^x = 1 + x + \frac{x^2}{2!} + \ldots + \frac{x^n}{n!} + O(x^{n+1}) \quad \text{as } x \to 0$$

and
$$f'(x) = \frac{f(x+h) - f(x)}{h} + o(h) \quad \text{as } h \to 0.$$

In numerical computation, we often generate a sequence of approximations which approach the desired result successively. We say that a sequence $\{x^k\}$ converges to x^* with the order $n(\geq 2)$ of convergence if
$$\lim_{k \to \infty} x^k = x^* \quad \text{and} \quad |x^{k+1} - x^*| = O(|x^k - x^*|^n).$$

For $n = 2$, it is called a **quadratically convergent** sequence. We say that a sequence $\{x^k\}$ converges to x^* linearly if there is a constant $C \in (0, 1)$ such that
$$\lim_{k \to \infty} \frac{|x^{k+1} - x^*|}{|x^k - x^*|} \leq C.$$

We say that a sequence $\{x^k\}$ converges to x^* superlinearly if
$$\lim_{k \to \infty} \frac{|x^{k+1} - x^*|}{|x^k - x^*|} = 0, \quad \text{i.e.,} \quad |x^{k+1} - x^*| = o(|x^k - x^*|).$$

A **superlinearly convergent** sequence converges faster than a **linearly convergent** sequence.

---
quadratically convergent sequence [2 次収束列]　　superlinearly convergent [超 1 次収束]　　linearly convergent [線形収束, 1 次収束]

1.4 Error propagation

Final results of computations of unknown quantities are approximations in general. They are not exact but involve errors. Such errors may result from a combination of the following effects. Round-off errors result from rounding and truncation errors result from truncating, as discussed above. Experimental errors are errors of given data, for instance, errors in measurements. It is very important to know how errors at the beginning and in later steps propagate in the computation and affect accuracy.

Theorem 1.1 *(Error Propagation)*

(i) *In addition and subtraction, an error bound for the results is given by the sum of the error bounds for the terms.*

(ii) *In multiplication and division, an approximate error bound for the relative error of the results is given by the sum of error bounds for the relative errors of the given numbers.*

Proof: Let $\hat{x}$ and $\hat{y}$ be approximation values of x and y, respectively. Then the errors of $\hat{x}$ and $\hat{y}$ are $e_1 = x - \hat{x}$ and $e_2 = y - \hat{y}$.

(i) Suppose that the error bounds for $\hat{x}$ and $\hat{y}$ are ϵ_1 and ϵ_2, that is,

$$|e_1| \leq \epsilon_1, \qquad |e_2| \leq \epsilon_2.$$

Then from

$$x + y = \hat{x} + \hat{y} + e_1 + e_2,$$

we get an error bound for the absolute error of $\hat{x} + \hat{y}$ by

$$|(x+y) - (\hat{x} + \hat{y})| = |e_1 + e_2| \leq |e_1| + |e_2| \leq \epsilon_1 + \epsilon_2.$$

The proof for subtraction is similar and is left to the reader.

(ii) The error bound for the relative errors are

$$\frac{|e_1|}{|x|} \leq \epsilon_{r1}, \qquad \frac{|e_2|}{|y|} \leq \epsilon_{r2}.$$

Then from

$$xy = \hat{x}\hat{y} + \hat{x}e_2 + \hat{y}e_1 + e_1 e_2,$$

---

error propagation [誤差伝播]　　experimental error [実験誤差]　　measurements [測量, 測定]　　accuracy [精度]

1.5 Significant digits

we get an error bound for the relative error of $\hat{x}\hat{y}$ as

$$\begin{aligned}\frac{|xy - \hat{x}\hat{y}|}{|xy|} &= \frac{|\hat{x}e_2 + \hat{y}e_1 + e_1 e_2|}{|xy|} \\ &= \frac{|(x - e_1)e_2 + (y - e_2)e_1 + e_1 e_2|}{|xy|} \\ &\approx \frac{|xe_2 + ye_1|}{|xy|} \\ &\leq \frac{|e_2|}{|y|} + \frac{|e_1|}{|x|} \\ &= \epsilon_{r1} + \epsilon_{r2}.\end{aligned}$$

Here, "approximately" means: we neglect $|e_1 e_2|$ as small compared to $|e_1|$ and $|e_2|$. The proof for division is left to the reader. ∎

1.5 Significant digits

Most digital computers have two ways of representing numbers, called fixed point and floating point. In a **fixed point** system, all numbers are represented with a fixed number of decimals, for example, 3.141, 0.002, 1000.000. In a **floating point** system, numbers are represented with a fixed number of significant digits, for example, $3.141 \times 10^0, 0.200 \times 10^{-2}, 1.000 \times 10^3$.

The fixed point form of numbers is impractical in engineering and science. We always use a floating point system, and significant digits.

A **significant digit** of a number x is any digit of x, except for zeros to the left of the first nonzero digit which serve only to fix the position of the decimal point. For instance, each of the numbers 62.4, 0.0123, 100 has 3 significant digits.

Consider the two numbers $x = 3.1415926$ and $y = 3.1415927$, which are nearly equal and both have 8 significant digits. However, their difference $x - y = 0.0000001$ has only 1 significant digit. This phenomenon is called loss of significant digits by substraction.

Poor computing methods can lead to loss of significant digits and give inaccurate results. It is important to choose the best method possible.

---

significant digit[有効けた]　　fixed point [固定小数点]　　floating point[浮動小数点]
loss of significant digits [けた落ち]

Even in very simple cases, one can use numerically good or numerically poor mathematics.

Let us find the roots of the quadratic equation

$$ax^2 + 2bx + c = 0$$

using 4 significant digits in the computation.

A formula for the roots x_1, x_2 is

$$x_1 = \frac{-b + \sqrt{b^2 - ac}}{a}, \quad x_2 = \frac{-b - \sqrt{b^2 - ac}}{a}. \tag{1.1}$$

Two other formulas for those roots are

$$x_1 = \frac{-c}{b + \sqrt{b^2 - ac}}, \quad x_2 = \frac{c}{\sqrt{b^2 - ac} - b} \tag{1.2}$$

and

$$x_1 = \frac{-b + \sqrt{b^2 - ac}}{a}, \quad x_2 = \frac{c}{ax_1}. \tag{1.3}$$

Let $a = 1, b = -20, c = 2$. We get

$$\begin{aligned} x_1 &= 39.95, & x_2 &= 0.05, & &\text{from (1.1)} \\ x_1 &= 40, & x_2 &= 0.05006, & &\text{from (1.2)} \\ x_1 &= 39.95, & x_2 &= 0.05006, & &\text{from (1.3).} \end{aligned}$$

Formulas (1.1) and (1.2) give poor results. Formula (1.3) is best, which gives results with 4 significant digits.

Remark: As we have seen, it is very important to estimate the errors for the final numerical results. In the last decades, a number of methods and software have been developed for verifying the numerical results given by computers. We recommend three books on numerical verification methods [Alefeld-Herzberger 1983], [Nakao-Yamamoto 1998] and [Oishi 2000].

---

estimate [評価]   verification methods [精度保証付方法]

Exercises

(1.1) Let a be an approximate value of x. Compute bounds for the absolute error and relative error of a.

(1.1.1) $\quad x = \pi, \quad a = 3.1415926$

(1.1.2) $\quad x = e, \quad a = \sum_{n=0}^{3} \frac{1}{n!}$

(1.1.3) $\quad x = \sqrt{3}, \quad a = 1.726$

(1.1.4) $\quad x = \sin\frac{1}{2}, \quad a = \frac{1}{2} - \frac{1}{6}\left(\frac{1}{2}\right)^3 + \frac{1}{120}\left(\frac{1}{2}\right)^5$

(1.1.5) $\quad x = \cos\frac{1}{2}, \quad a = 1 - \frac{1}{4}\left(\frac{1}{2}\right)^2 + \frac{1}{24}\left(\frac{1}{2}\right)^4$

(1.2) Using five digits, find approximate values of x in Exercise 1.1 with error at most 10^{-2}.

(1.3) Prove Theorem 1.1 (ii) for division.

(1.4) Solve $x^2 + 30x - 1 = 0$ by (1.1), (1.2) and (1.3), using 6 significant digits in the computation.

(1.5) Suppose that a sequence $\{a_k\}$ satisfies $a_k > 0$ and
$$1 > \frac{a_{n+2}}{a_{n+1}} \geq \frac{a_{n+3}}{a_{n+2}} \geq \cdots,$$
for a positive integer n. Let
$$S = \sum_{k=1}^{\infty} a_k \quad \text{and} \quad S_n = \sum_{k=1}^{n} a_k.$$
Show that the truncation error satisfies
$$S - S_n \leq \frac{a_{n+1}^2}{a_{n+1} - a_{n+2}}.$$

(1.6) Suppose that $|f^{(4)}| \leq M$. Show
$$f''(x) = \frac{f(x-h) - 2f(x) + f(x+h)}{h^2} + O(h^2).$$

Chapter 2.

Systems of Linear Equations

2.1 Introduction

A system of n linear equations in n unknowns $x_1, \ldots, x_n$ is of the form

$$
\begin{aligned}
a_{11}x_1 + \ldots + a_{1n}x_n &= b_1 \\
a_{21}x_1 + \ldots + a_{2n}x_n &= b_2 \\
&\ldots\ldots \\
a_{n1}x_1 + \ldots + a_{nn}x_n &= b_n,
\end{aligned}
$$

where the coefficients a_{ij} and the b_i are given numbers.

The system can be written as

$$\sum_{j=1}^{n} a_{ij}x_j = b_i, \quad i = 1, 2, \ldots, n.$$

Moreover, using matrix-vector notation, we can write the system as a single vector equation

$$Ax = b, \tag{2.1}$$

---

unknowns [未知数]   coefficients [係数]   matrix [行列] (複数形は matrices)

2.2 Gaussian elimination

where the coefficient matrix $A = [a_{ij}]$ is the $n \times n$ matrix

$$A = \begin{bmatrix} a_{11} & a_{12} & \cdots & a_{1n} \\ a_{21} & a_{22} & \cdots & a_{2n} \\ \vdots & \vdots & & \vdots \\ a_{n1} & a_{n2} & \cdots & a_{nn} \end{bmatrix}$$

and x and b are n dimensional vectors

$$x = \begin{bmatrix} x_1 \\ \vdots \\ x_n \end{bmatrix}, \quad b = \begin{bmatrix} b_1 \\ \vdots \\ b_n \end{bmatrix}.$$

A **solution** of (2.1) is a vector $x^* = [x_1^*, \ldots, x_n^*]^T$ which satisfies (2.1). In this chapter, we assume that A is nonsingular so the solution is unique

The system is called homogeneous if all b_i, $i = 1, 2, \ldots, n$ are zero. Obviously, a zero vector $x^* = [0, 0, \ldots, 0]^T$ is a solution of a homogeneous system.

Numerical methods for solving (2.1) are classified by two classes: **direct methods** and **iterative methods**. In this chapter, we shall study two direct methods which are the Gaussian elimination method and the LU-factorization, and two iterative methods which are the Jacobi method and the Gauss-Seidel method.

2.2 Gaussian elimination

The Gaussian elimination method for solving the linear system (2.1) is a systematic process of elimination that reduces (2.1) to **"triangular form"** because then the system can be easily solved by **"back substitution"**. For instance, a triangular system is

$$\begin{aligned} 3x_1 + 5x_2 + 2x_3 &= 0 \\ x_2 - 2x_3 &= 3 \\ 2x_3 &= 1 \end{aligned}$$

and back substitution gives $x_3 = 0.5$ from the third equation, then

$$x_2 = 3 + 2x_3 = 4$$

---

homogeneous [斉次、同次]　　direct method　　[直接法]　　iterative method [反復法]　　Gaussian elimination [ガウスの消去法]　　LU-factorization [LU 分解]　　Jacobi method [ヤコビ法]　　Gauss-Seidel method[ガウス・ザイデル法]　　triangular form [三角型]　　back substitution [後退代入]

from the second equation, and finally from the first equation

$$x_1 = \frac{1}{3}(0 - 5x_2 - 2x_3) = -7.$$

The Gaussian elimination method has two stages: **forward elimination** and **back substitution**

The forward elimination is a procedure to reduce a given system (2.1) to triangular form. Now we state the procedure.

Step 1 Suppose $a_{11} \neq 0$. We eliminate x_1 from the second equation to the nth equation in (2.1). This we do by adding the first equation multiplied by

$$m_i^{(1)} = -\frac{a_{i1}}{a_{11}}, \quad i = 2, \ldots, n$$

to the other equations. The resulting system is

$$\begin{cases} a_{11}x_1 + a_{12}x_2 + \ldots + a_{1n}x_n = b_1 \\ a_{22}^{(1)}x_2 + \ldots + a_{2n}^{(1)}x_n = b_2^{(1)} \\ \quad \ldots \ldots \\ a_{n2}^{(1)}x_2 + \ldots + a_{nn}^{(1)}x_n = b_n^{(1)}, \end{cases}$$

where

$$a_{ij}^{(1)} = a_{ij} + m_i^{(1)}a_{1j}, \quad b_i^{(1)} = b_i + m_i^{(1)}b_1, \quad 2 \leq i, j \leq n.$$

This first equation is called the **pivot equation** in this step, and a_{11} is called the **pivot**. This equation is left unaltered.

Step 2 We take the new second equation (which no longer contains x_1) as the pivot equation and use it to eliminate x_2 from the new third equation to the new nth equation. Let $a_{22}^{(1)} \neq 0$ and put

$$m_i^{(2)} = -\frac{a_{i2}^{(1)}}{a_{22}^{(1)}}, \quad a_{ij}^{(2)} = a_{ij}^{(1)} + m_i^{(2)}a_{2j}^{(1)}, \quad b_i^{(2)} = b_i^{(1)} + m_i^{(2)}b_2^{(1)},$$

$$3 \leq i, j \leq n.$$

$$\ldots \ldots$$

Step $n-1$ The $(n-1)$st equation is the pivot equation and $a_{n-1\,n-1}^{(n-2)}$ is the pivot. We use the pivot equation to eliminate x_{n-1} from the nth

---

forward elimination [前進消去]　　back substitution [後退代入]

2.2 Gaussian elimination

equation and get

$$\begin{cases} a_{11}x_1 + a_{12}x_2 \ldots + a_{1n}x_n = b_1 \\ a_{22}^{(1)}x_2 \ldots + a_{2n}^{(1)}x_n = b_2^{(1)} \\ \ldots\ldots \\ a_{nn}^{(n-1)}x_n = b_n^{(n-1)}. \end{cases}$$

Back substitution determine $x_n, x_{n-1}, \ldots, x_1$ from the upper triangular matrix obtained from Gaussian elimination. We have

$$x_n = \frac{b_n^{(n-1)}}{a_{nn}^{(n-1)}}$$

$$x_{n-1} = \frac{1}{a_{n-1\,n-1}^{(n-2)}}(b_{n-1}^{(n-2)} - a_{n-1\,n}^{(n-2)}x_n)$$

$$\ldots$$

$$x_1 = \frac{1}{a_{11}}(b_1 - a_{12}x_2 \ldots - a_{1n}x_n). \qquad \blacklozenge$$

The pivots must be different from zero. To achieve this we may have to change the order of equations. Also, pivots should not be too small in absolute value, because of round-off errors. Usually, the pivot at the ith step is chosen so that

$$|a_{ii}^{(i-1)}| = \max_{i \le k \le n} |a_{ki}^{(i-1)}|.$$

This is called **partial pivoting**.

Example 2.1 Apply the Gaussian elimination method with partial pivoting to solve the system

$$\begin{aligned} -x_2 + x_3 &= 4 \\ 2x_1 + 4x_2 + x_3 &= 3 \\ 3x_1 + 2x_2 - 2x_3 &= -2. \end{aligned}$$

Solution To get a pivot equation containing x_1, we have to reorder the equations by interchanging the first equation and the third equation which contains the maximum coefficient of x_1 in modulus. The resulting

---

partial pivoting [部分ピボット]

system is

$$3x_1 + 2x_2 - 2x_3 = -2$$
$$2x_1 + 4x_2 + x_3 = 3$$
$$-x_2 + x_3 = 4.$$

Step 1 Elimination of x_1 in the last two equations.
Compute
$$m_2^{(1)} = -\frac{2}{3}, \quad m_3^{(1)} = 0.$$

Add $m_2^{(1)}$ times the first equation to the second equation

$$a_{22}^{(1)} = 4 - \frac{2}{3} \times 2 = \frac{8}{3}, \quad a_{23}^{(1)} = 1 + \frac{2}{3} \times 2 = \frac{7}{3}, \quad b_2^{(1)} = 3 + \frac{2}{3} \times 2 = \frac{13}{3}.$$

Add $m_3^{(1)}$ times the first equation to the third equation

$$a_{32}^{(1)} = -1, \quad a_{33}^{(1)} = 1, \quad b_3^{(1)} = 4.$$

The resulting system is

$$3x_1 + 2x_2 - 2x_3 = -2$$
$$\frac{8}{3}x_2 + \frac{7}{3}x_3 = \frac{13}{3}$$
$$-x_2 + x_3 = 4.$$

Step 2 Elimination of x_2 in the last equation. Observe that in this step, it is not necessary to interchange the second and the third equations.
Compute
$$m_3^{(2)} = \frac{3}{8}.$$

Add $m_3^{(2)}$ times the new second equation to the third equation.

$$a_{33}^{(2)} = 1 + \frac{3}{8} \times \frac{7}{3} = \frac{15}{8}, \quad b_3^{(2)} = 4 + \frac{3}{8} \times \frac{13}{3} = \frac{45}{8}.$$

The resulting system is

$$3x_1 + 2x_2 - 2x_3 = -2$$
$$\frac{8}{3}x_2 + \frac{7}{3}x_3 = \frac{13}{3}$$
$$\frac{15}{8}x_3 = \frac{45}{8}.$$

2.2 Gaussian elimination

This is the end of the forward elimination. Now comes the back substitution.

Back substitution Determination of x_3, x_2, x_1.

$$x_3 = \frac{45}{15} = 3$$
$$x_2 = \frac{1}{8}(13 - 7 \times 3) = -1$$
$$x_1 = \frac{1}{3}(-2 + 2 \times 3 + 2) = 2. \qquad \blacklozenge\lozenge$$

In the forward elimination, we only use the matrix A and the vector b. Hence the forward elimination can be done on the **augmented matrix**

$$\tilde{A} = \begin{bmatrix} a_{11} & a_{12} & \cdots & a_{1n} & b_1 \\ a_{21} & a_{22} & \cdots & a_{2n} & b_2 \\ \vdots & \vdots & & \vdots & \vdots \\ a_{n1} & a_{n2} & \cdots & a_{nn} & b_n \end{bmatrix}$$

We illustrate the forward elimination on the augmented matrix of Example 2.1 as follows.

$$\begin{bmatrix} 0 & -1 & 1 & 4 \\ 2 & 4 & 1 & 3 \\ 3 & 2 & -2 & -2 \end{bmatrix} \qquad \begin{array}{l} row\ 1 \\ row\ 2 \\ row\ 3 \end{array}$$

Interchange row 1 and row 3:

$$\begin{bmatrix} 3 & 2 & -2 & -2 \\ 2 & 4 & 1 & 3 \\ & -1 & 1 & 4 \end{bmatrix} \qquad \begin{array}{l} row\ 1' \\ row\ 2' \\ row\ 3' \end{array}$$

row $1' \times (-2/3)$ + row $2'$:

$$\begin{bmatrix} 3 & 2 & -2 & -2 \\ & 8/3 & 7/3 & 13/3 \\ & -1 & 1 & 4 \end{bmatrix} \qquad \begin{array}{l} row\ 1'' \\ row\ 2'' \\ row\ 3'' \end{array}$$

row $2'' \times 3/8$ + row $3''$:

$$\begin{bmatrix} 3 & 2 & -2 & -2 \\ & 8/3 & 7/3 & 13/3 \\ & & 15/8 & 45/8 \end{bmatrix}.$$

---

augmented matrix [拡大行列] (n 次行列 A に n 次元ベクトル b をつけ加えてできる $n \times (n+1)$ 行列)

2.3 Matrix inversion

Gaussian elimination can be used for matrix inversion, where the situation is as follows.

Let e_j denote the vector whose jth component is one and other components are zero, that is, e_j is the jth column of the $n \times n$ identity matrix.

The inverse of a nonsingular $n \times n$ matrix A may be determined by solving the n systems

$$Ax = e_j, \quad j = 1, 2, \ldots, n.$$

That is, if we denote the solution of this system by $x^{(j)} = [x_1^{(j)}, \ldots, x_n^{(j)}]^T$, $j = 1, 2, \ldots, n$, then

$$A^{-1} = [x^{(1)}, \ldots, x^{(n)}] = \begin{bmatrix} x_1^{(1)} & \cdots & x_1^{(n)} \\ \vdots & & \vdots \\ x_n^{(1)} & \cdots & x_n^{(n)} \end{bmatrix}.$$

Example 2.2 Apply Gaussian elimination to find the inverse of the matrix in Example 2.1.

Solution The forward elimination can be down on the augmented matrix

$$\begin{bmatrix} 0 & -1 & 1 & 1 & 0 & 0 \\ 2 & 4 & 1 & 0 & 1 & 0 \\ 3 & 2 & -2 & 0 & 0 & 1 \end{bmatrix} \quad \begin{matrix} row\ 1 \\ row\ 2 \\ row\ 3 \end{matrix}$$

Interchange row 1 and row 3

$$\begin{bmatrix} 3 & 2 & -2 & 0 & 0 & 1 \\ 2 & 4 & 1 & 0 & 1 & 0 \\ & -1 & 1 & 1 & 0 & 0 \end{bmatrix} \quad \begin{matrix} row\ 1' \\ row\ 2' \\ row\ 3' \end{matrix}$$

row $1' \times (-2/3) +$ row $2'$:

$$\begin{bmatrix} 3 & 2 & -2 & 0 & 0 & 1 \\ & 8/3 & 7/3 & 0 & 1 & -2/3 \\ & -1 & 1 & 1 & 0 & 0 \end{bmatrix} \quad \begin{matrix} row\ 1'' \\ row\ 2'' \\ row\ 3'' \end{matrix}$$

row $2'' \times 3/8 +$ row $3''$:

$$\begin{bmatrix} 3 & 2 & -2 & 0 & 0 & 1 \\ & 8/3 & 7/3 & 0 & 1 & -2/3 \\ & & 15/8 & 1 & 3/8 & -2/8 \end{bmatrix} \quad \begin{matrix} row\ 1''' \\ row\ 2''' \\ row\ 3''' \end{matrix}$$

2.4 LU-factorization

We solve the three systems

$$\begin{bmatrix} 3 & 2 & -2 \\ & 8/3 & 7/3 \\ & & 15/8 \end{bmatrix} \begin{bmatrix} x_1 \\ x_2 \\ x_3 \end{bmatrix} = \begin{bmatrix} 0 \\ 0 \\ 1 \end{bmatrix},$$

$$\begin{bmatrix} 3 & 2 & -2 \\ & 8/3 & 7/3 \\ & & 15/8 \end{bmatrix} \begin{bmatrix} x_1 \\ x_2 \\ x_3 \end{bmatrix} = \begin{bmatrix} 0 \\ 1 \\ 3/8 \end{bmatrix},$$

$$\begin{bmatrix} 3 & 2 & -2 \\ & 8/3 & 7/3 \\ & & 15/8 \end{bmatrix} \begin{bmatrix} x_1 \\ x_2 \\ x_3 \end{bmatrix} = \begin{bmatrix} 1 \\ -2/3 \\ -2/8 \end{bmatrix}.$$

The solutions of the three systems are

$$x^{(1)} = \begin{bmatrix} 2/3 \\ -7/15 \\ 8/15 \end{bmatrix}, \quad x^{(2)} = \begin{bmatrix} 0 \\ 1/5 \\ 1/5 \end{bmatrix}, \quad x^{(3)} = \begin{bmatrix} 1/3 \\ -2/15 \\ -2/15 \end{bmatrix},$$

which form the inverse

$$A^{-1} = \begin{bmatrix} 2/3 & 0 & 1/3 \\ -7/15 & 1/5 & -2/15 \\ 8/15 & 1/5 & -2/15 \end{bmatrix}. \qquad \diamond$$

2.4 LU-factorization

An **LU-factorization** of a given square matrix A is of the form

$$A = LU,$$

where L is a **lower triangular** matrix and U is an **upper triangular** matrix. It can be proved that for any nonsingular matrix rows can be reordered so that the resulting matrix has an LU-factorization in which L turns out to be the matrix of the multipliers m_{ij} of Gaussian elimination, with main diagonal $1, \ldots, 1$, and U is the matrix of the triangular system at the end of Gaussian elimination.

The crucial idea is that L and U can be computed directly, without using Gaussian elimination. Once we have an LU-factorization of A, we can use it for solving $Ax = b$ in two steps

---

lower triangular [下三角]　　upper triangular [上三角]

$$\text{solving} \quad Ly = b, \quad \text{for } y$$
$$\text{solving} \quad Ux = y, \quad \text{for } x.$$

Example 2.3 Solve the following system by the LU factorization.

$$\begin{aligned} 3x_1 + 2x_2 - 2x_3 &= -2 \\ 2x_1 + 4x_2 + x_3 &= 3 \\ -x_2 + x_3 &= 4 \end{aligned}$$

Solution Let us set

$$A = \begin{bmatrix} 3 & 2 & -2 \\ 2 & 4 & 1 \\ 0 & -1 & 1 \end{bmatrix}, L = \begin{bmatrix} 1 & 0 & 0 \\ l_{21} & 1 & 0 \\ l_{31} & l_{32} & 1 \end{bmatrix}, U = \begin{bmatrix} u_{11} & u_{12} & u_{13} \\ 0 & u_{22} & u_{23} \\ 0 & 0 & u_{33} \end{bmatrix},$$

and $A = LU$. Then we have

$$\begin{bmatrix} 3 & 2 & -2 \\ 2 & 4 & 1 \\ 0 & -1 & 1 \end{bmatrix} = \begin{bmatrix} u_{11} & u_{12} & u_{13} \\ l_{21}u_{11} & l_{21}u_{12} + u_{22} & l_{21}u_{13} + u_{23} \\ l_{31}u_{11} & l_{31}u_{12} + l_{32}u_{22} & l_{31}u_{13} + l_{32}u_{23} + u_{33} \end{bmatrix}$$

Computing from the first row, we obtain

$u_{11} = 3, \quad u_{12} = 2, \quad u_{13} = -2$
$l_{21} = 2/u_{11} = 2/3, \quad u_{22} = 4 - l_{21}u_{12} = 8/3, \quad u_{23} = 1 - l_{21}u_{13} = 7/3$
$l_{31} = 0, l_{32} = (-1 - l_{31}u_{12})/u_{22} = -3/8, u_{33} = 1 - l_{31}u_{13} - l_{32}u_{23} = 15/8$

We first solve $Ly = b$, that is

$$\begin{bmatrix} 1 & 0 & 0 \\ 2/3 & 1 & 0 \\ 0 & -3/8 & 1 \end{bmatrix} \begin{bmatrix} y_1 \\ y_2 \\ y_3 \end{bmatrix} = \begin{bmatrix} -2 \\ 3 \\ 4 \end{bmatrix}. \quad \text{Solution} \quad y = \begin{bmatrix} -2 \\ 13/3 \\ 45/8 \end{bmatrix}.$$

Then we solve $Ux = y$, that is

$$\begin{bmatrix} 3 & 2 & -2 \\ 0 & 8/3 & 7/3 \\ 0 & 0 & 15/8 \end{bmatrix} \begin{bmatrix} x_1 \\ x_2 \\ x_3 \end{bmatrix} = \begin{bmatrix} -2 \\ 13/3 \\ 45/8 \end{bmatrix}. \quad \text{Solution} \quad x = \begin{bmatrix} 2 \\ -1 \\ 3 \end{bmatrix}.$$

◇

2.4 LU-factorization

If A is a symmetric positive definite matrix ($A = A^T$ and $x^T A x > 0$ for all $x \neq 0$), then there exists a unique real lower triangular matrix L with positive diagonals such that

$$A = LL^T.$$

A popular method of solving $Ax = b$ based on this factorization is called **Cholesky's method**. The formulas for the factorization are

$$l_{11} = \sqrt{a_{11}}, \qquad l_{i1} = \frac{a_{i1}}{l_{11}}, \qquad 2 \leq i \leq n$$

$$l_{jj} = \sqrt{a_{jj} - \sum_{k=1}^{j-1} l_{jk}^2}, \qquad j \geq 2$$

$$l_{ij} = \frac{1}{l_{jj}} \left(a_{ij} - \sum_{k=1}^{j-1} l_{ik} l_{jk} \right), \qquad j+1 \leq i \leq n.$$

Example 2.4 Solve the following system by Cholesky's method

$$\begin{aligned} 4x_1 + 2x_2 + 6x_3 &= 30 \\ 2x_1 + 17x_2 - x_3 &= 19 \\ 6x_1 - x_2 + 14x_3 &= 56 \end{aligned}$$

Solution From $A = LL^T$, we have

$$\begin{bmatrix} 4 & 2 & 6 \\ 2 & 17 & -1 \\ 6 & -1 & 14 \end{bmatrix} = \begin{bmatrix} l_{11} & 0 & 0 \\ l_{21} & l_{22} & 0 \\ l_{31} & l_{32} & l_{33} \end{bmatrix} \begin{bmatrix} l_{11} & l_{21} & l_{31} \\ 0 & l_{22} & l_{32} \\ 0 & 0 & l_{33} \end{bmatrix}.$$

Now we compute in the given order,

$$l_{11} = \sqrt{a_{11}} = 2, \qquad l_{21} = \frac{a_{21}}{l_{11}} = 1, \qquad l_{31} = \frac{a_{31}}{l_{11}} = 3$$

$$\begin{aligned} l_{22} &= \sqrt{a_{22} - l_{21}^2} = \sqrt{17-1} = 4 \\ l_{32} &= \frac{1}{l_{22}} (a_{32} - l_{31} l_{21}) = \frac{1}{4}(-1-3) = -1 \\ l_{33} &= \sqrt{a_{33} - l_{31}^2 - l_{32}^2} = \sqrt{14-9-1} = 2. \end{aligned}$$

---

symmetric positive definite matrix[対称正定値行列]　　Cholesky's method [コレスキー法]

We first solve $Ly = b$, that is

$$\begin{bmatrix} 2 & 0 & 0 \\ 1 & 4 & 0 \\ 3 & -1 & 2 \end{bmatrix} \begin{bmatrix} y_1 \\ y_2 \\ y_3 \end{bmatrix} = \begin{bmatrix} 30 \\ 19 \\ 56 \end{bmatrix}. \quad \text{Solution} \quad y = \begin{bmatrix} 15 \\ 1 \\ 6 \end{bmatrix}.$$

Then we solve $Ux = L^T x = y$, that is

$$\begin{bmatrix} 2 & 1 & 3 \\ 0 & 4 & -1 \\ 0 & 0 & 2 \end{bmatrix} \begin{bmatrix} x_1 \\ x_2 \\ x_3 \end{bmatrix} = \begin{bmatrix} 15 \\ 1 \\ 6 \end{bmatrix}. \quad \text{Solution} \quad x = \begin{bmatrix} 2.5 \\ 1 \\ 3 \end{bmatrix}. \quad \diamondsuit$$

2.5 Iterative methods

In an iterative method, we start from an approximation $x^0 = [x_1^0, \ldots, x_n^0]^T$ to the true solution $x^* = [x_1^*, \ldots, x_n^*]^T$, and, if successful, we obtain a sequence $\{x^k\}$, $x^k = [x_1^k, \ldots, x_n^k]^T$, which converges to x^*. For a given error bound $\epsilon > 0$, we usually stop the iterative method if

$$|x_i^{k+1} - x_i^k| \leq \epsilon, \quad i = 1, 2, \ldots n$$

or

$$|x_i^{k+1} - x_i^k| \leq \epsilon |x_i^k|, \quad i = 1, 2, \ldots n.$$

We assume that $a_{ii} \neq 0$ for $i = 1, 2, \ldots, n$. We now write

$$A = D + L + U,$$

where D is a diagonal matrix and L and U are respectively lower and upper triangular matrices with zero main diagonals. Thus $Ax = b$ can be written as

$$(D + L + U)x = b. \tag{2.2}$$

The Jacobi method is based on the equation

$$x = -D^{-1}(L + U)x + D^{-1}b,$$

and the Gauss-Seidel method is based on the equation

$$x = -(L + D)^{-1}Ux + (L + D)^{-1}b.$$

---

diagonal matrix [対角行列]　　lower triangular matrix [下三角行列]　　upper triangular matrix [上三角行列]

2.5 Iterative methods

Jacobi method

$$x^{k+1} = -D^{-1}(L+U)x^k + D^{-1}b, \quad \text{for } k \geq 0.$$

The Jacobi method in component form is

$$x_i^{k+1} = \frac{1}{a_{ii}} \left(b_i - \sum_{\substack{j=1 \\ j \neq i}}^{n} a_{ij} x_j^k \right), \quad i = 1, 2, \ldots, n, \quad k \geq 0.$$

Gauss-Seidel method

$$x^{k+1} = -(L+D)^{-1} U x^k + (L+D)^{-1} b, \quad \text{for } k \geq 0.$$

The Gauss-Seidel method in components form is

$$x_i^{k+1} = \frac{1}{a_{ii}} \left(b_i - \sum_{i<j}^{n} a_{ij} x_j^k - \sum_{i>j}^{n} a_{ij} x_j^{k+1} \right), \quad i = 1, 2, \ldots, n, \quad k \geq 0.$$

Example 2.5 Apply the Jacobi method and Gauss-Seidel method to solve the system

$$\begin{aligned} 10x_1 + x_2 + x_3 &= 12 \\ 2x_1 + 10x_2 + x_3 &= 13 \\ 2x_1 + 2x_2 + 10x_3 &= 14 \end{aligned}$$

starting from $x^0 = [0, 0, 0]^T$. Write down the first two steps of the methods.

Solution We rewrite the system as

$$\begin{aligned} x_1 &= \frac{12 - x_2 - x_3}{10} \\ x_2 &= \frac{13 - 2x_1 - x_3}{10} \\ x_3 &= \frac{14 - 2x_1 - 2x_2}{10} \end{aligned}$$

Jacobi method:
Step 1

$$x_1^1 = \frac{12 - x_2^0 - x_3^0}{10} = 1.2$$
$$x_2^1 = \frac{13 - 2x_1^0 - x_3^0}{10} = 1.3$$
$$x_3^1 = \frac{14 - 2x_1^0 - 2x_2^0}{10} = 1.4$$

Step 2

$$x_1^2 = \frac{12 - x_2^1 - x_3^1}{10} = \frac{12 - 1.3 - 1.4}{10} = 0.93$$
$$x_2^2 = \frac{13 - 2x_1^1 - x_3^1}{10} = \frac{13 - 2 \times 1.2 - 1.4}{10} = 0.92$$
$$x_3^2 = \frac{14 - 2x_1^1 - 2x_2^1}{10} = \frac{14 - 2 \times 1.2 - 2 \times 1.3}{10} = 0.9. \quad \blacklozenge$$

Gauss-Seidel method:
Step 1

$$x_1^1 = \frac{12 - x_2^0 - x_3^0}{10} = 1.2$$
$$x_2^1 = \frac{13 - 2x_1^1 - x_3^0}{10} = \frac{13 - 2 \times 1.2 - 0}{10} = 1.06$$
$$x_3^1 = \frac{14 - 2x_1^1 - 2x_2^1}{10} = \frac{14 - 2 \times 1.2 - 2 \times 1.06}{10} = 0.948$$

Step 2

$$x_1^2 = \frac{12 - x_2^1 - x_3^1}{10} = \frac{12 - 1.06 - 0.948}{10} = 0.999$$
$$x_2^2 = \frac{13 - 2x_1^2 - x_3^1}{10} = \frac{13 - 2 \times 0.999 - 0.948}{10} = 1.005$$
$$x_3^2 = \frac{14 - 2x_1^2 - 2x_2^2}{10} = \frac{14 - 2 \times 0.999 - 2 \times 1.005}{10} = 0.999.$$

(The true solution is $x^* = [1, 1, 1]^T$.) $\blacklozenge\lozenge$

2.6 Vector norms and matrix norms

The numerical analysis of iterative methods requires the use of vector norms and matrix norms in analyzing their convergence properties.

Vector norm A vector norm is a real valued function $\|\cdot\| : C^n \to R$, which satisfies the following three properties for all $x, y \in C^n$.

(VN1) $\|x\| = 0 \Leftrightarrow x = 0$, and $\|x\| \geq 0$ (Positivity).

(VN2) $\|\alpha x\| = |\alpha| \|x\|$ for all numbers α (Homogeneity).

(VN3) $\|x + y\| \leq \|x\| + \|y\|$ (Triangle inequality).

It follows from (VN2) and (VN3) that

$$\|x + y\| \geq \|x\| - \|y\|,$$

since $\|-y\| = \|y\|$ and

$$\|x + y\| = \|x + y\| + \|-y\| - \|y\| \geq \|(x+y) + (-y)\| - \|y\| = \|x\| - \|y\|.$$

The value $\|x\|$ is called a vector norm of $x \in C^n$. The norm of a vector provides a measure of the magnitude of the vector, and it can be used to define the distance between two vectors in C^n.

A sequence $\{x^k\} \subset C^n$ is said to converge to the vector $x^* \in C^n$ if

$$\lim_{k \to \infty} \|x^k - x^*\| = 0$$

for some vector norm $\|\cdot\|$. The vector x^* is then called the limit of the sequence $\{x^k\}$; we write

$$x^* = \lim_{k \to \infty} x^k.$$

Note that the limit of a convergent sequence is unique. Furthermore, it can be shown that in the finite-dimensional space, the definition of convergence does not depend on the choice of a vector norm. That is, if $\{x^k\}$ converges to x^* for some norm $\|\cdot\|$, then $\{x^k\}$ converges to x^* for any other norm.

The most important norm in connection with computation is the p-norm defined by

$$\|x\|_p = (|x_1|^p + |x_2|^p + \ldots + |x_n|^p)^{1/p},$$

---

vector norm [ベクトルノルム]   matrix norm [行列ノルム]   positivity [正値性]   homogeneity [斉次性]   triangle inequality [三角不等式]   convergent sequence [収束列]

where p is a fixed number and $p \geq 1$. In practice, one usually takes $p = 1$, $p = 2$, or $p = \infty$, and labels them by a subscript:

$$\|x\|_1 = |x_1| + \ldots + |x_n| \qquad l_1 \text{ - norm}$$
$$\|x\|_2 = \sqrt{|x_1|^2 + \ldots + |x_n|^2} \qquad \text{Euclidean norm or } l_2\text{- norm}$$
$$\|x\|_\infty = \max_j |x_j| \qquad l_\infty\text{- norm.}$$

Example 2.6 Compute the l_1, l_2, l_∞ norms of the vectors $x = [-1, -3, 4]$ and $y = [1, -1, 1, -1]$.

Solution

$$\begin{aligned}
\|x\|_1 &= |x_1| + |x_2| + |x_3| = |-1| + |-3| + |4| = 8. \\
\|x\|_2 &= \sqrt{x_1^2 + x_2^2 + x_3^2} = \sqrt{1 + 9 + 16} = \sqrt{26}. \\
\|x\|_\infty &= \max\{|x_1|, |x_2|, |x_3|\} = \max\{1, 3, 4\} = 4.
\end{aligned}$$

$$\begin{aligned}
\|y\|_1 &= |y_1| + |y_2| + |y_3| + |y_4| = |1| + |-1| + |1| + |-1| = 4. \\
\|y\|_2 &= \sqrt{y_1^2 + y_2^2 + y_3^2 + y_4^2} = \sqrt{1 + 1 + 1 + 1} = \sqrt{4} = 2. \\
\|y\|_\infty &= \max\{|y_1|, |y_2|, |y_3|, |y_4|\} = \max\{1, 1, 1, 1\} = 1. \quad \diamond
\end{aligned}$$

Matrix norms Let $C^{n \times n}$ be the set of $n \times n$ matrices with complex elements. A matrix norm is a real valued function $\|\cdot\| : C^{n \times n} \to R$ which satisfies the following four properties for all $A, B \in C^{n \times n}$.

(MN1) $\|A\| = 0 \Leftrightarrow A = 0$, and $\|A\| \geq 0$ (Positivity).

(MN2) $\|\alpha A\| = |\alpha| \|A\|$ for all numbers α (Homogeneity).

(MN3) $\|A + B\| \leq \|A\| + \|B\|$ (Triangle inequality).

(MN4) $\|AB\| \leq \|A\| \|B\|$ (Submultiplicativity).

It follows from (MN2) and (MN3) that

$$\|A + B\| \geq \|A\| - \|B\|,$$

since $\|-B\| = \|B\|$ and

$$\|A+B\| = \|A+B\| + \|-B\| - \|B\| \geq \|(A+B)+(-B)\| - \|B\| = \|A\| - \|B\|.$$

2.6 Vector norms and matrix norms

The value $\|A\|$ is called a matrix norm of A. For an n-dimensional vector x, Ax is an n-dimensional vector. We now take a vector norm and consider $\|x\|$ and $\|Ax\|$. One can prove that there is a number c such that

$$\|Ax\| \le c\|x\| \qquad \text{for all } x.$$

Let $x \ne 0$. Then $\|x\| > 0$ by (VN1) and division gives

$$\frac{\|Ax\|}{\|x\|} \le c.$$

This implies

$$\sup_{x \ne 0} \frac{\|Ax\|}{\|x\|} \le c < \infty.$$

If we define a functional $\|\cdot\| : C^{n \times n} \to R$ by

$$\|A\| = \sup_{x \ne 0} \frac{\|Ax\|}{\|x\|},$$

then we can show that the function satisfies the four properties of a matrix norm. This function is called the **matrix norm induced by the vector norm**. Alternatively, it can be written as

$$\|A\| = \max_{\|x\|=1} \|Ax\|.$$

For the matrix norm induced by the vector norm, we have an additional property
(MN5) $\|Ax\| \le \|A\|\|x\|$ for all x.

Note that $\|A\|$ depends on the vector norm that we picked. In particular, for the norm $\|A\|_p$ induced by the vector norm $\|x\|_p$, we have

$$\|A\|_p = \sup_{x \ne 0} \frac{\|Ax\|_p}{\|x\|_p}.$$

Specially for $p = 1, 2, \infty$, we can show that

$$\|A\|_1 = \max_j \sum_{i=1}^n |a_{ij}| \qquad l_1\text{ - norm}$$
$$\|A\|_2 = \sqrt{\text{the largest eigenvalue of } A^T A} \qquad l_2\text{ - norm}$$
$$\|A\|_\infty = \max_i \sum_{j=1}^n |a_{ij}| \qquad l_\infty\text{ - norm}.$$

---

functional [汎関数] (実数値をとる写像を汎 (はん) 関数という)　　matrix norm induced by the vector norm [ベクトルノルムに誘導される行列ノルム]

See Exercise (2.10).

Since $\|A\|_2$ is difficult to compute in general, the Euclidean norm

$$\|A\|_E = \sqrt{\sum_{i,j=1}^{n} |a_{ij}|^2}$$

is often used in place of $\|A\|_2$. It satisfies

$$\|A\|_2 \leq \|A\|_E \leq \sqrt{n}\|A\|_2.$$

Condition number of a matrix   The condition number $\kappa(A)$ of a nonsingular matrix A is defined by

$$\kappa(A) = \|A\|\|A^{-1}\|.$$

It should be noted that for any matrix norm induced by a vector norm, we have $\kappa(A) \geq 1$ since $\|A\|\|A^{-1}\| \geq \|AA^{-1}\| = \|I\| = 1$.

In error analysis of linear equations, it is very impotant to know how variations in the matrix A and the vector b affect the solution. The following theorem shows that if the condition number $\kappa(A)$ is small, then small perturbations introduced in A and b will not cause big errors in the solution. In this case, A is called a well-conditioned matrix, and the system of linear equations is called a well-conditioned system.

Theorem 2.1   Let $Ax = b$, $(A + \triangle A)(x + \triangle x) = b + \triangle b$,

$$e_x = \frac{\|\triangle x\|}{\|x\|}, \qquad e_A = \frac{\|\triangle A\|}{\|A\|} \quad \text{and} \quad e_b = \frac{\|\triangle b\|}{\|b\|}.$$

Suppose A is nonsingular, $b \neq 0$ and $\|A^{-1}\|\|\triangle A\| < 1$. Then, there holds

$$e_x \leq \frac{\kappa(A)}{1 - \kappa(A)e_A}(e_A + e_b).$$

Proof: From the assumptions, we have

$$\begin{aligned}A^{-1}(b + \triangle b) &= A^{-1}(A + \triangle A)(x + \triangle x) \\ &= (I + A^{-1}\triangle A)x + (I + A^{-1}\triangle A)\triangle x.\end{aligned}$$

---

condition number of a matrix [行列の条件数]   nonsingular matrix [正則行列]   perturbation [摂動]

2.6 Vector norms and matrix norms

Using $A^{-1}b = x$, we simplify this as

$$(I + A^{-1}\triangle A)\triangle x = A^{-1}\triangle b - (A^{-1}\triangle A)x.$$

Now by $\|A^{-1}\|\|\triangle A\| < 1$, we get

$$\begin{aligned}\|(I + A^{-1}\triangle A)\triangle x\| &\geq (1 - \|A^{-1}\triangle A\|)\|\triangle x\| \\ &\geq (1 - \|A^{-1}\|\|\triangle A\|)\|\triangle x\| \\ &= (1 - \kappa(A)e_A)\|\triangle x\|,\end{aligned}$$

and

$$\begin{aligned}\|(I + A^{-1}\triangle A)\triangle x\| &\leq \|A^{-1}\triangle b\| + \|(A^{-1}\triangle A)x\| \\ &\leq \|A^{-1}\|\|A\|\left(\frac{\|\triangle b\|}{\|A\|\|x\|} + \frac{\|\triangle A\|}{\|A\|}\right)\|x\| \\ &\leq \kappa(A)(e_b + e_A)\|x\|.\end{aligned}$$

Piecing together the two inequalities

$$(1 - \kappa(A)e_A)\|\triangle x\| \leq \|(I + A^{-1}\triangle A)\triangle x\| \leq \kappa(A)(e_b + e_A)\|x\|,$$

we obtain the assertion of the theorem. ∎

Remark If the condition number $\kappa(A)$ is large, then A is called an ill-conditioned matrix. Theorem 2.1 shows that if A is ill-conditioned, then small perturbations in A and b may cause big errors in the solution, that is, the system is ill-conditioned. Note that a well-conditioned system can have an ill-conditioned coefficient matrix. See an example in [Yamamoto 2003 (page 64)].

Example 2.7 Compute the matrix norms and condition numbers corresponding to the l_1, l_2, l_∞ norms for

$$A = \begin{bmatrix} 2 & 1 \\ 0 & 1 \end{bmatrix}.$$

Solution

$$\begin{aligned}\|A\|_1 &= \max(2+0, 1+1) = 2 \\ \|A\|_\infty &= \max(2+1, 0+1) = 3\end{aligned}$$

To compute l_2 norm, we have to find the largest eigenvalue of

$$A^T A = \begin{bmatrix} 4 & 2 \\ 2 & 2 \end{bmatrix}.$$

The characteristic equation of the matrix is

$$\det(A^T A - \lambda I) = (4 - \lambda)(2 - \lambda) - 4 = \lambda^2 - 6\lambda + 4 = 0.$$

Hence $A^T A$ has two eigenvalues

$$\lambda_1 = 3 + \sqrt{5}, \quad \lambda_2 = 3 - \sqrt{5}.$$

Thus

$$\|A\|_2 = \sqrt{\lambda_1}.$$

To compute the condition number, we have to know the inverse of A, which is

$$A^{-1} = \begin{bmatrix} 0.5 & -0.5 \\ 0 & 1 \end{bmatrix}.$$

It is also easy to see $(A^{-1})^T A^{-1}$ has two eigenvalues $1/\lambda_1, 1/\lambda_2$ and $\|A^{-1}\|_2 = 1/\sqrt{\lambda_2}$.

We then have

$$\begin{aligned} \kappa_1(A) &= \|A\|_1 \|A^{-1}\|_1 = 2 \times \max(0.5, 1.5) = 3 \\ \kappa_\infty(A) &= \|A\|_\infty \|A^{-1}\|_\infty = 3 \times \max(1, 1) = 3 \\ \kappa_2(A) &= \|A\|_2 \|A^{-1}\|_2 = \sqrt{\lambda_1/\lambda_2} = \frac{1}{2}(3 + \sqrt{5}) \approx 2.6180. \end{aligned} \qquad \diamondsuit$$

Spectral radius The spectral radius $\rho(A)$ of an $n \times n$ matrix A is defined by

$$\rho(A) = \max_{1 \leq i \leq n} |\lambda_i|,$$

where $\lambda_i, i = 1, 2, \ldots, n$ are eigenvalues of A.

The spectral radius is closely related to the norm of a matrix, as is shown in the following theorem.

Theorem 2.2 *For any matrix norm $\|A\|$ of A induced by the vector norm, we have*

$$\rho(A) \leq \|A\|.$$

Conversely, for any $\epsilon > 0$, there exists a matrix norm such that

$$\|A\| \leq \rho(A) + \epsilon.$$

---

spectral radius [スペクトル半径]

2.7 Convergence of iterative methods

Proof: Let λ be an eigenvalue of A and x be an eigenvector of A corresponding to λ. Since $Ax = \lambda x$ and $\|x\| \neq 0$, we have

$$|\lambda|\|x\| = \|\lambda x\| = \|Ax\| \leq \|A\|\|x\|,$$

where we use the matrix norm property (MN5). Thus

$$\rho(A) = \max_{1 \leq i \leq n} |\lambda_i| \leq \|A\|.$$

See [Yamamoto 2003, Theorem 3.3], for the proof of the second part of this theorem. ∎

2.7 Convergence of iterative methods

Most iterative methods for solving the system of linear equations (2.1) can be written as

$$x^{k+1} = Hx^k + c, \qquad k \geq 0 \qquad (2.3)$$

where H is an $n \times n$ matrix. In fact, if we split A into $A = M - N$, where M is nonsingular, then the system $Ax = b$ is equivalent to

$$Mx = Nx + b$$

or

$$x = M^{-1}Nx + M^{-1}b.$$

Hence, we can define an iterative method (2.3) with $H = M^{-1}N$ and $c = M^{-1}b$.

The Jacobi method and Gauss-Seidel method are special cases of (2.3). In the Jacobi method, $H = -D^{-1}(L + U)$ and $c = D^{-1}b$. In the Gauss-Seidel method, $H = -(D + L)^{-1}U$ and $c = (D + L)^{-1}b$.

Theorem 2.3 *For any $c \in C^n$, the equation $x = Hx + c$ has a unique solution and the sequence $\{x^k\}$ defined by (2.3) converges to the unique solution from any starting point $x^0 \in C^n$ if and only if $\rho(H) < 1$.*

Proof: If $\rho(H) < 1$, then every eigenvalue of H satisfies $|\lambda| < 1$ and the matrix $I - H$ is nonsingular. Thus the system

$$(I - H)x = c$$

has a unique solution x^*.

From (2.3),
$$x^k - x^* = H(x^{k-1} - x^*) = H^k(x^0 - x^*).$$

By Theorem 2.2, there is a matrix norm $\|\cdot\|$ such that $\|H\| < 1$. This implies that
$$\lim_{k\to\infty} \|H^k\| \leq \lim_{k\to\infty} \|H\|^k = 0,$$
so that
$$\lim_{k\to\infty} H^k = 0.$$
Hence the sequence $\{x^k\}$ converges to x^* as
$$\lim_{k\to\infty} \|x^k - x^*\| = \lim_{k\to\infty} \|H^k(x^0 - x^*)\| = 0.$$

Conversely, suppose that the sequence $\{x^k\}$ defined by (2.3) converges to the unique solution x^* of $x = Hx + c$ from any starting point x^0. Let λ be an eigenvalue of H and let v be an eigenvector of H corresponding to λ. Then the iterations (2.3) with $x^0 = x^* + v$ satisfies
$$x^k - x^* = H^k(x^0 - x^*) = H^k v = \lambda^k v.$$
Hence
$$|\lambda^k|\|v\| = \|x^k - x^*\| \to 0 \quad \text{as } k \to \infty.$$
This implies $|\lambda|^k \to 0$ as $k \to \infty$ since $\|v\| > 0$. Consequently we obtain $|\lambda| < 1$ or $\rho(H) < 1$. ∎

Example 2.8 For the system of linear equations with a coefficient matrix
$$A = \begin{bmatrix} 3 & 2 & -2 \\ 2 & 4 & 1 \\ 0 & -1 & 1 \end{bmatrix},$$
determine whether the Jacobi method and the Guass-Seidel method are convergent or divergent.

Solution The iterative matrix in the Jacobi method is
$H_J = -D^{-1}(L+U)$

$$= -\begin{bmatrix} 3 & 0 & 0 \\ 0 & 4 & 0 \\ 0 & 0 & 1 \end{bmatrix}^{-1} \begin{bmatrix} 0 & 2 & -2 \\ 2 & 0 & 1 \\ 0 & -1 & 0 \end{bmatrix} = \begin{bmatrix} 0 & -2/3 & 2/3 \\ -1/2 & 0 & -1/4 \\ 0 & 1 & 0 \end{bmatrix}.$$

---

convergent [収束]   divergent[発散]

2.7 Convergence of iterative methods

Solving the characteristic equation

$$\det(H_J - \lambda I) = -\lambda^3 + \frac{1}{12}\lambda - \frac{1}{3} = 0,$$

we find the eigenvalues of H_J

$$\lambda_1 = -0.7334, \quad \lambda_2 = 0.3667 + 0.5657i, \quad \lambda_3 = 0.3667 - 0.5657i.$$

Hence the spectral radius $\rho(H_J) = 0.7334$. By Theorem 2.3, the Jacobi method is convergent.

The iterative matrix in the Gauss-Seidel method is
$H_G = -(D+L)^{-1}U$

$$= -\begin{bmatrix} 3 & 0 & 0 \\ 2 & 4 & 0 \\ 0 & -1 & 1 \end{bmatrix}^{-1} \begin{bmatrix} 0 & 2 & -2 \\ 0 & 0 & 1 \\ 0 & 0 & 0 \end{bmatrix} = \begin{bmatrix} 0 & -2/3 & 2/3 \\ 0 & 1/3 & -7/12 \\ 0 & 1/3 & -7/12 \end{bmatrix}.$$

Solving the characteristic equation

$$\det(H_G - \lambda I) = -\lambda\left(\left(\frac{1}{3}-\lambda\right)\left(-\frac{7}{12}-\lambda\right) + \frac{7}{36}\right) = -\lambda^2\left(\lambda + \frac{1}{4}\right) = 0$$

we find the eigenvalues of H_G

$$\lambda_1 = -0.25, \quad \lambda_2 = \lambda_3 = 0.$$

Hence the spectral radius $\rho(H_G) = 0.25$. By Theorem 2.3, the Gauss-Seidel method is convergent. ◇

Moreover, from $\rho(H_G) = 0.25 < \rho(H_J) = 0.7334$, we can claim that the Gauss-Seidel method for solving the linear equations with the coefficient matrix A in Example 2.8 converges faster than the Jacobi method.

If the spectral radius of the iterative matrix is known, then we know whether the method is convergent or divergent. However, calculating the spectral radius is in general more difficult than solving the linear system.

Now we give some sufficient conditions for convergence of the Jacobi and Gauss-Seidel methods, which are easily verified.

The following corollary follows from Theorems 2.2 and 2.3.

Corollary 2.1 *If $\|H\| < 1$ for a matrix norm induced by the vector norm, then the sequence $\{x^k\}$ defined by (2.3) converges to the unique solution of $x = Hx + c$ from any starting point x^0.*

We say A is strictly diagonally dominant if

$$|a_{ii}| > \sum_{\substack{j=1 \\ j \neq i}}^{n} |a_{ij}|, \quad i = 1, 2, \ldots, n.$$

Theorem 2.4 *If A is strictly diagonally dominant, then for any starting point x^0, both the Jacobi and Gauss-Seidel methods generate sequences $\{x^k\}$ that converge to the unique solution of $Ax = b$.*

Proof: It is easy to show the convergence of the Jacobi method, because

$$\|D^{-1}(L+U)\|_\infty = \max_i \left(\frac{1}{|a_{ii}|} \sum_{\substack{j=1 \\ j \neq i}}^{n} |a_{ij}| \right) < 1.$$

Now we shall show the convergence of the Gauss-Seidel method. Let $y \in R^n$ be an arbitrary vector with $\|y\|_\infty = 1$. Consider the equation

$$(D+L)x = -Uy.$$

By induction, we show

$$|x|_i = |(D+L)^{-1} Uy|_i \leq \left| \frac{1}{a_{ii}} \right| \sum_{\substack{j=1 \\ j \neq i}}^{n} |a_{ij}| < 1, \quad \text{for } i = 1, 2, \ldots, n. \quad (2.4)$$

For $i = 1$, we have

$$|x_1| = \left| -\frac{1}{a_{11}} \sum_{j=2}^{n} a_{1j} y_j \right| \leq \left| \frac{1}{a_{11}} \right| \sum_{j=2}^{n} |a_{1j}||y_j| \leq \left| \frac{1}{a_{11}} \right| \sum_{j=2}^{n} |a_{1j}| < 1.$$

Suppose that

$$|x_i| \leq \left| \frac{1}{a_{ii}} \right| \sum_{\substack{j=1 \\ j \neq i}}^{n} |a_{ij}| < 1, \quad i = 1, \ldots, m-1 \quad (m \leq n).$$

---

strictly diagonally dominant[狭義優対角]

2.7 Convergence of iterative methods

Then we get

$$\begin{aligned}
|x_m| &= \left|\frac{1}{a_{mm}}\right| \left(\left|\sum_{j=1}^{m-1} a_{mj}x_j + \sum_{j=m+1}^{n} a_{mj}y_j\right|\right) \\
&\leq \left|\frac{1}{a_{mm}}\right| \left(\sum_{j=1}^{m-1} |a_{mj}||x_j| + \sum_{j=m+1}^{n} |a_{mj}||y_j|\right) \\
&\leq \left|\frac{1}{a_{mm}}\right| \sum_{\substack{j=1\\j\neq m}}^{n} |a_{mj}| \\
&< 1.
\end{aligned}$$

This implies that (2.4) holds, and hence

$$\|(D+L)^{-1}Uy\|_\infty < 1.$$

Since y is an arbitrary vector with $\|y\|_\infty = 1$, we have

$$\|(D+L)^{-1}U\|_\infty = \max_{\|y\|_\infty=1} \|(D+L)^{-1}Uy\|_\infty < 1.$$

The convergence now follows from Corollary 2.1. ∎

Example 2.9 Show that the Jacobi method and the Gauss-Seidel method are convergent for the system of linear equations with a coefficient matrix

$$A = \begin{bmatrix} 4 & -1 & 0 \\ -1 & 4 & -1 \\ 0 & -1 & 6 \end{bmatrix}$$

Solution The coefficient matrix A is strictly diagonally dominant, since

$$|a_{11}| = 4 > |a_{12}| + |a_{13}| = 1$$
$$|a_{22}| = 4 > |a_{21}| + |a_{23}| = 2$$
$$|a_{33}| = 6 > |a_{31}| + |a_{32}| = 1.$$

The convergence of the two methods follows from Theorem 2.4. ◇

Exercises

(2.1) Apply the Gaussian elimination method and LU factorization to solve the system
$$\begin{aligned} 2x_1 - x_2 &= 1 \\ -x_1 + 2x_2 - x_3 &= 0 \\ -x_2 + 2x_3 &= 1 \end{aligned}$$

(2.2) Apply the Jacobi method and the Gauss-Seidel method to solve the system
$$\begin{bmatrix} 5 & -2 & 2 \\ -1 & 3 & -1 \\ -2 & -2 & 6 \end{bmatrix} \begin{bmatrix} x_1 \\ x_2 \\ x_3 \end{bmatrix} = \begin{bmatrix} 1 \\ -1 \\ -3 \end{bmatrix}$$
starting from $x^0 = (1, 1, 1)$. Write down the first three steps.

(2.3) Apply Cholesky's method to solve the system
$$\begin{bmatrix} 4 & -2 & 0 \\ -2 & 2 & -1 \\ 0 & -1 & 2 \end{bmatrix} \begin{bmatrix} x_1 \\ x_2 \\ x_3 \end{bmatrix} = \begin{bmatrix} 2 \\ 0 \\ -1 \end{bmatrix}$$

(2.4) Compute the vector norms $\|x\|_1, \|x\|_2, \|x\|_\infty, \|y\|_1, \|y\|_2, \|y\|_\infty$ for the vectors
$$x = [1, -2, 0, -8, 2, 7, 3, 4]$$
$$y = [-1, 3, 6, 4, 5, -7, -9, 3].$$
Show that $\|\lambda x\|_p = |\lambda| \|x\|_p$ and $\|x + y\|_p \leq \|x\|_p + \|y\|_p$ hold for any real number λ and $p = 1, 2, \infty$.

(2.5) Compute the matrix norms $\|A\|_1, \|A\|_2, \|A\|_\infty, \|B\|_1, \|B\|_2, \|B\|_\infty$ for the matrices
$$A = \begin{bmatrix} -2 & 1 & 0 \\ -1 & 2 & -1 \\ 0 & -1 & 2 \end{bmatrix}, \quad B = \begin{bmatrix} 4 & -2 & 0 \\ -2 & 2 & -1 \\ 0 & -1 & 2 \end{bmatrix}.$$
Show that $\|\lambda A\| = |\lambda| \|A\|$, $\|A + B\| \leq \|A\| + \|B\|$ and $\|AB\| \leq \|A\| \|B\|$ hold for any real number λ and $p = 1, 2, \infty$.

(2.6) Compute the spectral radii $\rho(A), \rho(B)$ of matrices A, B in Exercise (2.5). Verify $\rho(A) \leq \|A\|_p$ and $\rho(B) \leq \|B\|_p$ for $p = 1, 2, \infty$.

(2.7) Compute the condition number $\kappa_p(A) = \|A\|_p \|A^{-1}\|_p$ of the matrix A in Exercise (2.5) for $p = 1, 2, \infty$.

(2.8) Do the Jacobi method and the Gauss-Seidel method converge to the unique solution of the system in Exercise (2.1) from any starting point?

(2.9) Let x be an n-dimensional vector. Show that $\|x\|_1, \|x\|_2, \|x\|_\infty$ satisfy conditions (VN1), (VN2) and (VN3).

(2.10) Let A be an $n \times n$ matrix. Show that $\|A\|_1, \|A\|_2, \|A\|_\infty$ satisfy conditions (MN1), (MN2), (MN3), (MN4).

(2.11) Show that
$$\lim_{p \to \infty} \|x\|_p = \|x\|_\infty.$$

(2.12) Show that from any starting point the Jacobi method converges to the unique solution of the following system
$$\begin{bmatrix} 1 & -2 & 2 \\ -1 & 1 & -1 \\ -2 & -2 & 1 \end{bmatrix} \begin{bmatrix} x_1 \\ x_2 \\ x_3 \end{bmatrix} = \begin{bmatrix} 1 \\ -1 \\ -3 \end{bmatrix}$$
but the Gauss-Seidel method does not.

(2.13) Show that from any starting point the Gauss-Seidel method converges to the unique solution of the following system
$$\begin{bmatrix} 2 & 1 & 1 \\ 2 & 2 & 2 \\ 1 & 1 & 2 \end{bmatrix} \begin{bmatrix} x_1 \\ x_2 \\ x_3 \end{bmatrix} = \begin{bmatrix} 1 \\ -1 \\ -3 \end{bmatrix}$$
but the Jacobi method does not.

(2.14) Show that the Jacobi method and the Gauss-Seidel method converge to the unique solution of the system in Exercise (2.2) from any starting point.

Chapter 3.

Systems of Nonlinear Equations

3.1 Introduction

We study numerical methods for finding solutions of a system of nonlinear equations

$$f(x) = 0, \tag{3.1}$$

where f is a continuously differentiable function from R^n into itself and $x \in R^n$. If $n = 1$, then (3.1) reduces to a single equation.

A **solution** of (3.1) is a vector x^* such that $f(x^*) = 0$.

For $n = 1$, if f is a polynomial, then (3.1) is called an **algebraic equation**, and its solutions are called **roots** of the equation. If $n = 1$ and f involves transcendental functions, (3.1) is called a transcendental equation.

There are practically no formulas (except in a few simple cases) for solving (3.1), so one depends almost entirely on iterative methods.

In an iterative method, we start from an initial point $x_0 \in R^n$ and compute step by step approximations $x_1, x_2, \ldots$ to an unknown solution of (3.1). We hope that the sequence $\{x_k\}$ converges to a solution of (3.1). (We say $\{x_k\}$ converges to x^* if $\lim_{k \to \infty} x_k = x^*$).

Usually, the iteration is terminated when one of the following criteria holds:

(i) $\|x_{k+1} - x_k\| \leq \epsilon$

---

polynomial [多項式]　　algebraic equation [代数方程式]　　roots of the equation [方程式の根]　　transcendental function [超越関数]

3.2 The fixed point method

(ii) $\|x_{k+1} - x_k\| \leq \epsilon \|x_k\|$
(iii) $\|f(x_k)\| \leq \epsilon$.

Here ϵ is a given small positive number, which is called a stopping constant. Note that none of these three criteria can ensure

$$\|x_k - x^*\| \leq \epsilon.$$

We begin by studying the case of a single nonlinear equation with one unknown. In the first three sections, we consider the fixed point method, the Newton method and the secant method for solving a single equation in R, the set of real numbers. In Section 3.5, we extend the Newton method to R^n. In Section 3.6, we study how to find all zeros of a polynomial in C, the set of complex numbers.

3.2 The fixed point method

Let $f : R \to R$ and we reformulate (3.1) into the form

$$x = g(x). \tag{3.2}$$

Then we choose an x_0 and compute

$$x_{k+1} = g(x_k), \quad k = 0, 1, \ldots. \tag{3.3}$$

This is called a fixed point method. A solution of (3.2) is called a fixed point of g, which is a solution of (3.1). From (3.1), we may get several different forms of (3.2), and the behavior of corresponding iterative sequences $x_0, x_1, \ldots$ may differ accordingly.

Example 3.1 Find an approximate solution of

$$f(x) = x^2 - 3x + 2 = 0$$

using a fixed point method.

Since we know the equation has two solutions

$$x^* = 1, \qquad x^{**} = 2,$$

we can watch the behavior of different fixed point methods as the iteration proceeds.

---

fixed point method [不動点法]

Solution The equation can be written

$$x = g(x) = \frac{1}{3}(x^2 + 2).$$

Thus we can define a fixed point method by

$$x_{k+1} = \frac{1}{3}(x_k^2 + 2). \tag{3.4}$$

The equation may also be written

$$x = g(x) = \frac{x^2 - 2}{2x - 3}.$$

Thus we can define other fixed point method

$$x_{k+1} = \frac{x_k^2 - 2}{2x_k - 3}. \tag{3.5}$$

Table 3.1 gives numerical results of the two methods for various starting points x_0.

Table 3.1: Example 3.1

		$x_{k+1} = (x_k^2 + 2)/3$		
x_0	0.0	1.5	1.6	3.0
x_1	0.6667	1.4167	1.5200	3.6667
x_2	0.8148	1.3356	1.4368	5.1481
x_3	0.8880	1.2613	1.3548	9.5011
x_4	0.9295	1.1970	1.2785	30.7572
		$x_{k+1} = (x_k^2 - 2)/(2x_k - 3)$		
x_0	0.0	1.5	1.6	3.0
x_1	0.6667	-	2.8000	2.3333
x_2	0.9333	-	2.2462	2.0667
x_3	0.9961	-	2.0406	2.0039
x_4	1.0000	-	2.0015	2.0000

If we choose $x_0 = 0$, both methods seem to approach the solution $x^* = 1$. If we choose $x_0 = 1.5$, the situation in (3.4) is similar, but (3.5) fails. If we choose $x_0 = 1.6$, (3.4) seems to approach the solution $x^* = 1$, while (3.5) seems to approach the solution $x^* = 2$. If we choose $x_0 = 3.0$, (3.4) seems to diverge, but method (3.5) seems to approach the solution $x^* = 2$.

The behavior of the two fixed point methods can be explained by Figure 3.1. ◇

A sufficient condition for convergence of fixed point methods is given in the following theorem, which has various practical applications.

3.2 The fixed point method

$$g(x) = \frac{x^2+2}{3}$$

$$g(x) = \frac{x^2-2}{2x-3}$$

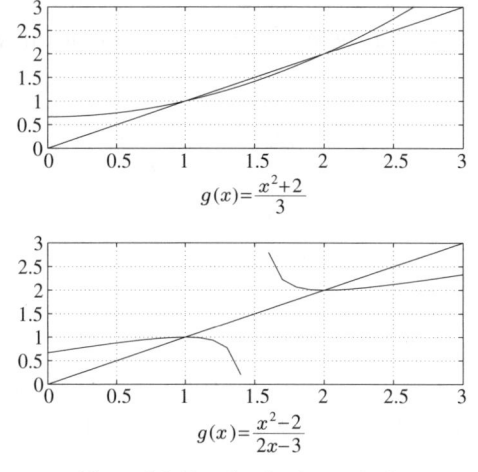

Figure 3.1: Two fixed point methods in Example 3.1

Theorem 3.1 *(Convergence of fixed point methods)* Let x^* be a solution of (3.2). Suppose that g has a continuous derivative with $|g'(x)| \leq \lambda < 1$ in an interval $I = [x^* - d, x^* + d] (d > 0)$. Then

(i) the iteration defined by (3.3) converges to x^* for any x_0 in I.
(ii) x^* is the unique solution of (3.2) in I.
(iii) x_k satisfies

$$\frac{\epsilon_k}{1+\lambda} \leq |x_k - x^*| \leq \frac{\epsilon_k}{1-\lambda} \leq \frac{\lambda^k \epsilon_0}{1-\lambda},$$

where $\epsilon_k = |x_{k+1} - x_k|$.

Proof: (i) Let $x \in I$. By the mean value theorem of differential calculus, there is t between x and x^* such that

$$g(x) - g(x^*) = g'(t)(x - x^*).$$

Hence

$$\begin{aligned} |g(x) - x^*| &= |g(x) - g(x^*)| \\ &\leq \lambda |x - x^*| \\ &\leq \lambda d \\ &< d. \end{aligned}$$

This implies that $g(x) \in I$. Since $x_{k+1} = g(x_k)$, the whole sequence $\{x_k\}$ remains in I. Using the mean value theorem again, we obtain

$$\begin{aligned} |x_k - x^*| &= |g(x_{k-1}) - g(x^*)| \\ &= |g'(t_{k-1})||x_{k-1} - x^*| \\ &\leq \lambda |x_{k-1} - x^*| \\ &= \lambda |g(x_{k-2}) - g(x^*)| \\ &= \lambda |g'(t_{k-2})||x_{k-2} - x^*| \\ &\leq \lambda^2 |x_{k-2} - x^*| \\ &\leq \lambda^k |x_0 - x^*|, \end{aligned}$$

where t_{k-1} is between x_{k-1} and x^*, and t_{k-2} is between x_{k-2} and x^*. Since $0 \leq \lambda < 1$, we have $\lambda^k \to 0$, as $k \to 0$. Thus $|x_k - x^*| \to 0$ as $k \to \infty$.

(ii) Assume that there is another solution $\hat{x}$. Then

$$0 < |x^* - \hat{x}| = |g(x^*) - g(\hat{x})| \leq \lambda |x^* - \hat{x}| < |x^* - \hat{x}|.$$

This is a contradiction. Hence x^* is the unique solution of (3.2) in I.

(iii) The first inequality follows from

$$\begin{aligned} \epsilon_k &= |x_{k+1} - x_k| \\ &\leq |x_{k+1} - x^*| + |x_k - x^*| \\ &\leq |g(x_k) - g(x^*)| + |x_k - x^*| \\ &\leq (\lambda + 1)|x_k - x^*|. \end{aligned}$$

The second inequality follows from

$$\begin{aligned} |x_k - x^*| &\leq |x_k - x_{k+1}| + |g(x_k) - x^*| \\ &\leq \epsilon_k + \lambda |x_k - x^*|. \end{aligned}$$

Finally, the third inequality follows from

$$\epsilon_k = |x_{k+1} - x_k| = |g(x_k) - g(x_{k-1})| \leq \lambda |x_k - x_{k-1}| = \lambda \epsilon_{k-1} \leq \ldots \leq \lambda^k \epsilon_0.$$

∎

Remark The proof of Theorem 3.1 means that if I is an interval and a function $g : I \to I$ satisfies the Lipschitz condition

$$|g(x) - g(y)| \leq \lambda |x - y|, \quad \text{for } x, y \in I$$

3.3 Newton's method

with
$$\lambda \in [0, 1),$$
then g has a fixed point in I and the point can be obtained by the iteration (3.3). Such a function g is called a contraction mapping, the number λ is called a contraction number and the result is called a contraction mapping principle.

In Example 3.1, the numerical results of iteration (3.5) exhibit a very rapid convergence. This is due to the fact that because of

$$g'(x) = \left(\frac{x^2 - 2}{2x - 3}\right)' = \frac{2x^2 - 6x + 4}{(2x - 3)^2} = 0$$

at the fixed points $x^* = 1$ and $x^* = 2$, the contraction number is very small. We shall show in the next section that iteration (3.5) is derived from the Newton method.

3.3 Newton's method

The Newton method is a popular iterative method for solving (3.1), because of its simplicity and great speed.

Newton's method is defined by

$$x_{k+1} = x_k - \frac{f(x_k)}{f'(x_k)}, \quad k = 0, 1, 2, \ldots.$$

There are many ways of introducing Newton's method. One way is based on Taylor polynomials. Suppose that f is twice continuously differentiable. Let x_k be an approximation to x^* such that $f'(x_k) \neq 0$ and $|x^* - x_k|$ is sufficiently small. Consider the Taylor polynomial for $f(x)$ expanded about x_k,

$$f(x) = f(x_k) + (x - x_k)f'(x_k) + \frac{1}{2}(x - x_k)^2 f''(\xi),$$

where ξ lies between x_k and x. Suppose $\tilde{x}$ is a good approximation of x^*. This equation, with $x = \tilde{x}$, gives

$$f(\tilde{x}) = f(x_k) + (\tilde{x} - x_k)f'(x_k) + \frac{1}{2}(\tilde{x} - x_k)^2 f''(\tilde{\xi}),$$

---

contraction mapping [縮小写像]　　contraction number [縮小数]　　contraction mapping principle [縮小写像の原理]　　Newton's method [ニュートン法]

where $\tilde{\xi}$ lies between x_k and $\tilde{x}$. Since we assume that $|x^* - x_k|$ is sufficiently small and $\tilde{x}$ is a good approximation of x^*, the term involving $(\tilde{x} - x_k)^2$ is negligible and

$$0 \approx f(\tilde{x}) \approx f(x_k) + (\tilde{x} - x_k)f'(x_k).$$

Solving

$$f(x_k) + (\tilde{x} - x_k)f'(x_k) = 0$$

for $\tilde{x}$, we obtain

$$\tilde{x} = x_k - \frac{f(x_k)}{f'(x_k)}.$$

Setting $x_{k+1} = \tilde{x}$, we get the Newton method. Geometrically, x_{k+1} stands for the intersection of the axis with the tangent of f at x_k.

In Example 3.1, iteration (3.5) is the Newton method applied to $f(x) = x^2 - 3x + 2$. In fact, we have $f'(x_k) = 2x_k - 3$ and

$$x_{k+1} = x_k - \frac{x_k^2 - 3x_k + 2}{2x_k - 3} = \frac{x_k^2 - 2}{2x_k - 3}.$$

Example 3.2 Set up a Newton iteration for computing the square root $x = \sqrt{2}$.

Solution Consider

$$f(x) = x^2 - 2 = 0.$$

Then $f'(x) = 2x$. Applying the Newton method to $f(x) = 0$ gives

$$x_{k+1} = x_k - \frac{x_k^2 - 2}{2x_k} = \frac{1}{2}\left(x_k + \frac{2}{x_k}\right).$$

If we choose $x_0 = 1$, we get

$$x_1 = 1.500, \qquad x_2 = 1.417, \qquad x_3 = 1.414, \qquad \ldots. \qquad \diamondsuit$$

Example 3.3 Apply Theorem 3.1 to the Newton method for solving $f(x) = x^2 - 2 = 0$, and show that for any starting point $x_0 \in I = [\sqrt{2} - 0.5, \sqrt{2} + 0.5]$, the Newton method converges to $\sqrt{2}$.

Solution As was seen in Example 3.2, the Newton method for finding $\sqrt{2}$ can be written as a fixed point method

$$x = g(x) = \frac{1}{2}\left(x + \frac{2}{x}\right).$$

The derivative
$$g'(x) = \frac{1}{2}\left(1 - \frac{2}{x^2}\right)$$
is continuous and strictly increasing in the interval I. Hence the maximum value of $|g'(x)|$ is at the end-points of the interval. Since
$$g'(\sqrt{2} - \frac{1}{2}) = -\frac{1}{2} \cdot \frac{\sqrt{2} - 1/4}{2 - \sqrt{2} + 1/4} > -\frac{3}{4}$$
and
$$g'(\sqrt{2} + \frac{1}{2}) = \frac{1}{2} \cdot \frac{\sqrt{2} + 1/4}{2 + \sqrt{2} + 1/4} < \frac{1}{2},$$
we have $|g'(x)| \leq 3/4$ in the interval I. By Theorem 3.1, for any starting point $x_0 \in I = [\sqrt{2} - 0.5, \sqrt{2} + 0.5]$, the Newton method for solving $f(x) = x^2 - 2 = 0$ converges to the solution $\sqrt{2}$. $\diamond$

In the above example, if we observe that $g(x) > \sqrt{2}$ for $x \in (0, \sqrt{2})$ and $\sqrt{2} < g(x) < x$ for $x > \sqrt{2}$, then, for any $x_0 > 0$, we have $x_1 \geq x_2 \geq \ldots \geq x_k \geq \sqrt{2}$. This implies a more precise result that the Newton method converges to $\sqrt{2}$, starting from any point $x_0 > 0$.

Remark Newton's method is extremely powerful and widely used, but needs the value of the derivative $f'(x)$ at each step. The steps used to compute $f(x)$ are usually easy to modify to obtain the required value of $f'(x)$ as well, see [Rall 1969, Chap.4, section 24, Kubota and Iri 1998 and Oishi 1999]. This technique is called automatic differentiation, and is not difficult to carry out by hand for a function of a single variable.

3.4 The secant method

As a variant of the Newton method, the **secant method**

$$x_{k+1} = x_k - f(x_k)\frac{x_k - x_{k-1}}{f(x_k) - f(x_{k-1})}, \quad k = 0, 1, 2, \cdots \tag{3.6}$$

is known, which replaces each derivative $f'(x_k)$ in the Newton method by difference quotient

$$\frac{f(x_k) - f(x_{k-1})}{x_k - x_{k-1}}.$$

Geometrically, this means that the tangent line at x_k is replaced by the secant line through the two points $(x_k, f(x_k))$ and $(x_{k-1}, f(x_{k-1}))$. We observe that two starting values x_0 and x_{-1} are necessary in the secant method and that this method is not quite as fast as Newton's method.

It is not good to write (3.6) as

$$x_{k+1} = \frac{x_{k-1} f(x_k) - x_k f(x_{k-1})}{f(x_k) - f(x_{k-1})},$$

because this may lead to loss of significant digits if x_k and x_{k-1} are about equal. Of course, the second term in (3.6) is also subject to loss of significant digits under the same condition, but the effect is usually not so great.

Figure 3.2 shows the difference between the Newton method and the secant method for solving $f(x) = x^2 - 2 = 0$ when we take $x_0 = 3$ and $x_{-1} = 1$. The Newton method generates an approximate solution $x_1 = 11/6$ which is the zero of the linear equation

$$f(3) + f'(3)(x - 3) = 7 + 6(x - 3) = 6x - 11 = 0.$$

The secant method generates an approximate solution $x_1 = 5/4$ which is the zero of the linear equation

$$f(3) + \frac{f(3) - f(1)}{3 - 1}(x - 3) = 7 + 4(x - 3) = 4x - 5 = 0.$$

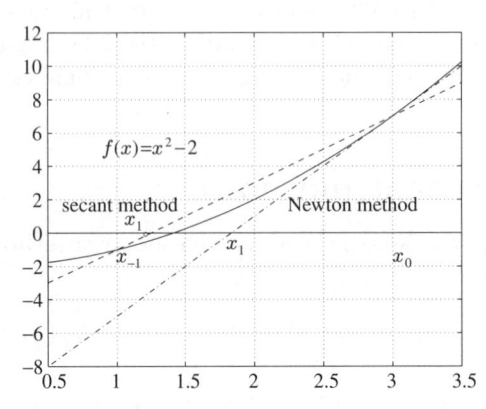

Figure 3.2: The Newton method and the secant method for $f(x)=x^2-2=0$.

Example 3.4 Solve $f(x) = x - \sin x = 0$ by the secant method, starting from $x_{-1} = 1.0$ and $x_0 = 0.9$, and $x_{-1} = 2.0$ and $x_0 = 1.0$.

Since we know $x = 0$ is a solution, we can watch the behavior of the iteration.

Solution With the definition of $f(x)$, (3.6) takes the form

$$x_{k+1} = x_k - \frac{(x_k - \sin x_k)(x_k - x_{k-1})}{x_k - x_{k-1} + (\sin x_{k-1} - \sin x_k)}.$$

From $x_{-1} = 1.0$ and $x_0 = 0.9$, we obtain

$$x_1 \approx 0.6213, \ x_2 \approx 0.4802, \ x_3 \approx 0.3574, \ x_4 \approx 0.2705\ldots$$

From $x_{-1} = 2.0$ and $x_0 = 1.0$, we obtain

$$x_1 \approx 0.8299, \ x_2 \approx 0.5945, \ x_3 \approx 0.4540, \qquad x_4 \approx 0.3396, \ldots. \qquad \diamondsuit$$

3.5 Systems of nonlinear equations

We now extend the Newton method to the case of systems of nonlinear equations. Consider the system of two nonlinear equations

$$f_1(x, y) = x^2 + 4y^2 - 1 = 0 \qquad (3.7)$$

$$f_2(x, y) = 2x^3 - x - 1 - 2y = 0. \qquad (3.8)$$

Solutions of this system are four points where the curves defined by $f_1(x,y) = 0$ and $f_2(x,y) = 0$ cross. The curves are

$$y = \pm\frac{1}{2}\sqrt{1 - x^2} \quad \text{and} \quad y = x^3 - \frac{1}{2}(x+1).$$

Figure 3.3 shows that the system of (3.7)-(3.8) has four solutions

$$(x_1^*, y_1^*) = \left(0, -\frac{1}{2}\right), \qquad (x_2^*, y_2^*) = (1, 0)$$

$$(x_3^*, y_3^*) \approx (-0.58, -0.41), \qquad (x_4^*, y_4^*) \approx (0.9, -0.22).$$

To extend the Newton method to a system of nonlinear equations with several variables, we have to generalize the "derivative" to the Jacobian.

Assume that $f_1(x,y)$ and $f_2(x,y)$ are functions of the independent variables x and y. Then the Jacobian matrix of

$$f(x, y) = \begin{bmatrix} f_1(x, y) \\ f_2(x, y) \end{bmatrix}$$

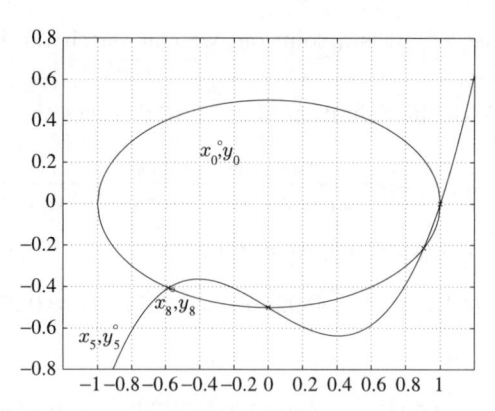

Figure 3.3: The Newton method in Example 3.5

is defined by
$$f'(x,y) = \begin{bmatrix} \dfrac{\partial f_1(x,y)}{\partial x} & \dfrac{\partial f_1(x,y)}{\partial y} \\ \dfrac{\partial f_2(x,y)}{\partial x} & \dfrac{\partial f_2(x,y)}{\partial y} \end{bmatrix}.$$

The Newton method is then defined by
$$\begin{bmatrix} x^{(k+1)} \\ y^{(k+1)} \end{bmatrix} = \begin{bmatrix} x^{(k)} \\ y^{(k)} \end{bmatrix} - f'(x^{(k)}, y^{(k)})^{-1} f(x^{(k)}, y^{(k)}).$$

Instead of inverting the Jacobian matrix $f'(x^{(k)}, y^{(k)})$, one simply solves the linear system of equations
$$f'(x^{(k)}, y^{(k)}) \left[\begin{bmatrix} x^{(k+1)} \\ y^{(k+1)} \end{bmatrix} - \begin{bmatrix} x^{(k)} \\ y^{(k)} \end{bmatrix} \right] = -f(x^{(k)}, y^{(k)})$$

for the differences
$$h_1^{(k)} = x^{(k+1)} - x^{(k)}, \quad h_2^{(k)} = y^{(k+1)} - y^{(k)}$$

and set
$$x^{(k+1)} = x^{(k)} + h_1^{(k)}, \quad y^{(k+1)} = y^{(k)} + h_2^{(k)}.$$

Example 3.5 Use the Newton method to solve the system of nonlinear equations (3.7)-(3.8), starting from $(x^{(0)}, y^{(0)}) = (-0.3, 0.3)$.

---

system of nonlinear equations [連立非線形方程式]

3.5 System of nonlinear equations

Solution The Jacobian matrix of f is
$$f'(x,y) = \begin{bmatrix} 2x & 8y \\ 6x^2 - 1 & -2 \end{bmatrix}.$$

To get $[x^{(1)}, y^{(1)}]^T$, we solve the linear system
$$\begin{bmatrix} -2 \times 0.3 & 8 \times 0.3 \\ 6 \times 0.09 - 1 & -2 \end{bmatrix} \begin{bmatrix} h_1^{(0)} \\ h_2^{(0)} \end{bmatrix} = - \begin{bmatrix} 0.09 + 4 \times 0.09 - 1 \\ -2 \times 0.027 + 0.3 - 1 - 0.6 \end{bmatrix}$$

and find that
$$\begin{bmatrix} h_1^{(0)} \\ h_2^{(0)} \end{bmatrix} = \begin{bmatrix} -1.8878 \\ -0.2428 \end{bmatrix}.$$

Hence we obtain
$$\begin{bmatrix} x^{(1)} \\ y^{(1)} \end{bmatrix} = \begin{bmatrix} x^{(0)} \\ y^{(0)} \end{bmatrix} + \begin{bmatrix} h_1^{(0)} \\ h_2^{(0)} \end{bmatrix} = \begin{bmatrix} -0.3 \\ 0.3 \end{bmatrix} + \begin{bmatrix} -1.8878 \\ -0.2428 \end{bmatrix}$$
$$= \begin{bmatrix} -2.1878 \\ 0.0572 \end{bmatrix}.$$

Similarly, we compute $(x^{(2)}, y^{(2)}), (x^{(3)}, y^{(3)}) \ldots$ In Table 3.2, we list some iterates. We can see that after 9 iterations, we obtain an approximate solution $(-0.5825, -0.4064)$. $\diamondsuit$

Table 3.2: Example 3.5

k	0	1	3	5	6
$x^{(k)}$	−0.3	−2.1878	−1.4613	−0.9091	−0.7123
$y^{(k)}$	0.3	0.0572	−2.2794	−0.5891	−0.4073
k	7	8	9	10	
$x^{(k)}$	−0.6092	−0.5841	−0.5825	−0.5825	
$y^{(k)}$	−0.3999	−0.4061	−0.4064	−0.4064	

For a system of n equations with n unknonws, we generally write
$$x = [x_1, x_2, \ldots, x_n]^T$$

$$f(x) = \begin{bmatrix} f_1(x_1, x_2, \ldots, x_n) \\ \vdots \\ f_n(x_1, x_2, \ldots, x_n) \end{bmatrix} = 0. \tag{3.9}$$

The Jacobian of f is the $n \times n$ matrix

$$f'(x) = \begin{bmatrix} \dfrac{\partial f_1(x)}{\partial x_1} & \cdots & \dfrac{\partial f_1(x)}{\partial x_n} \\ \vdots & & \vdots \\ \dfrac{\partial f_n(x)}{\partial x_1} & \cdots & \dfrac{\partial f_n(x)}{\partial x_n} \end{bmatrix}.$$

The Newton method is then written

$$x^{(k+1)} = x^{(k)} - f'(x^{(k)})^{-1} f(x^{(k)}),$$

where $x^{(k)} = [x_1^k, \ldots, x_n^k]^T$ and $x^{(0)}$ is appropriately given as a starting vector.

The Newton method is usully performed as the following procedure:

For $k = 0, 1, \ldots$

(N1) Solve the linear system
$$f'(x^{(k)}) h^{(k)} = -f(x^{(k)}) \quad \text{for } h^{(k)},$$

(N2) Put $x^{(k+1)} = x^{(k)} + h^{(k)}$,

(N3) Increase k to $k+1$ and return to (N1).

In some cases, the Jacobian matrices needed can be calculated by automatic differentiation (see [Rall 1969], [Kubota-Iri 1998] or [Oishi 1999]). Alternatively, at the price of slower convergence, we can also replace the Jacobian matrix $f'(x^{(k)})$ in the Newton method by a matrix A_k, which satisfies the secant equation

$$A_k (x^{(k)} - x^{(k-1)}) = f(x^{(k)}) - f(x^{(k-1)}).$$

The resulting method is called the secant method or the quasi-Newton method (see [Ortega-Rheinboldt 1970]).

Concerning the convergence of the Newton method, the following result is known [Yamamoto 2003].

Theorem 3.2 *Let $x^* = [x_1^*, \ldots, x_n^*]^T$ be a solution of (3.9) in a domain $D \subset R^n$. Assume that the second partial derivatives*

$$\frac{\partial^2 f_i(x)}{\partial x_i \partial x_j}, \quad i, j = 1, 2, \ldots, n$$

exist and are continuous in D for every i, j. If the $n \times n$ matrix $f'(x^*)$ is nonsingular and $x^{(0)}$ is sufficiently close to x^*, then the Newton method

---

second partial derivatives [２階偏導関数、偏微係数]

3.6 Zeros of Polynomials

starting from $x^{(0)}$ converges to x^* and satisfies

$$\|x^{(k+1)} - x^*\| \leq K\|x^{(k)} - x^*\|^2, \quad k = 0, 1, 2, \ldots, \tag{3.10}$$

where $K > 0$ is a constant.

The right-hand side of (3.10) may simply be written as $O(\|x^{(k)} - x^*\|^2)$. The convergence property (3.10) is called "quadratic convergence". This is the distinguishing feature of Newton's method. The method can also be applied to solve nonlinear operator equations in abstract spaces (see [Rall 1969]).

3.6 Zeros of polynomials

In this section, we study the Weierstrass method for finding the zeros of polynomials $p(z)$ with real coefficients. We know that a polynomial of degree n

$$p(z) = z^n + a_1 z^{n-1} + \ldots + a_n, \qquad a_n \neq 0$$

has n zeros, $z_1^*, z_2^*, \ldots, z_n^*$, where each zero is counted repeatedly according to its multiplicity.

Let

$$w(z) = \prod_{j=1}^{n}(z - z_j).$$

Then

$$w'(z) = \sum_{i=1}^{n} \prod_{j \neq i}^{n}(z - z_j)$$

and

$$w'(z_i) = \prod_{j \neq i}^{n}(z_i - z_j).$$

Hence, as an analogy of Newton's method applied to a single equation

$$p(z) = \prod_{j=1}^{n}(z - z_j^*) = 0,$$

the following method is proposed under the assumption that $z_1^*, \ldots, z_n^*$ are distinct.

---

quadratic convergence [2次収束]   zeros of polynomials [多項式の零点]

Starting from $z^{(0)} = (z_1^0, z_2^0, \ldots, z_n^0)$ with n complex numbers $z_1^0, z_2^0, \ldots, z_n^0$, the sequence $z^{(k)} = (z_1^k, z_2^k, \ldots, z_n^k), k = 1, 2, \ldots$ is generated by

$$z_i^{k+1} = z_i^k - \frac{p(z_i^k)}{\prod_{j \neq i}(z_i^k - z_j^k)}, \quad i = 1, 2, \ldots, n, \quad k \geq 0. \tag{3.11}$$

This method was proposed independently by Weierstrass(1891), Durand(1960), Dochev(1962), etc. Weierstrass used the method to prove the existence of zeros of an algebraic equation. Dochev first proved the convergence of (3.11) under the assumption that $z^{(0)}$ is sufficiently close to $z^* = (z_1^*, \ldots, z_n^*)$ and Kerner(1966) proved that (3.11) is the Newton method applied to the system of nonlinear equations

$$f_i(z) = (-1)^i \phi_i(z) - a_i = 0, \quad i = 1, 2, \ldots, n,$$

where ϕ_i denote the ith elementary symmetric functions:

$$\phi_i(z) = \sum_{1 \leq j_1 < \cdots < j_i \leq n} z_{j_1} z_{j_2} \ldots z_{j_i},$$

that is,

$$\begin{aligned}\phi_1(z) &= z_1 + z_2 + \ldots + z_n \\ \phi_2(z) &= z_1 z_2 + z_1 z_3 + \ldots + z_{n-1} z_n \\ &\vdots \\ \phi_n(z) &= z_1 z_2 \ldots z_n.\end{aligned}$$

Hence, the convergence speed of (3.11) is quadratic if $z^{(0)}$ is a good approximation to z^*. The method is called by different names such as the Weierstrass method, Durand-Kerner method, Weierstrass-Dochev method, etc.

Example 3.6 Use the Weierstrass method to find zeros of the polynomial

$$p(z) = ((z-1)^2 + 1)(z-2) = z^3 - 4z^2 + 6z - 4.$$

Solution We start from $z_1^0 = 3 - 3i, z_2^0 = 3 + 3i, z_3^0 = 3$. Applying the Weierstrass method to $p(z)$, we get

$$z_1^{k+1} = z_1^k - \frac{p(z_1^k)}{(z_1^k - z_2^k)(z_1^k - z_3^k)}$$

---

elementary symmetric function [基本対称関数](σ を $\{1, 2, \ldots, n\}$ の置換とする。任意の σ に対して $\phi(z_1, \ldots, z_n) = \phi(z_{\sigma_1}, \ldots, z_{\sigma_n})$ を満たす関数 ϕ を $z_1, \ldots z_n$ に関する対称関数という。$\phi_1, \ldots, \phi_n$ は根と係数の関係を表すそれぞれ 1 次,$\ldots$, n 次の対称式であり、これらを基本対称式という)

3.6 Zeros of Polynomials

$$z_2^{k+1} = z_2^k - \frac{p(z_2^k)}{(z_2^k - z_1^k)(z_2^k - z_3^k)}$$

$$z_3^{k+1} = z_3^k - \frac{p(z_3^k)}{(z_3^k - z_1^k)(z_3^k - z_2^k)}.$$

The computed results are listed in Table 3.3. ◇

Table 3.3: Example 3.6

k	z_1^k	z_2^k	z_3^k
0	3.0000 − 3.0000i	3.0000 + 3.0000i	3.0000 + 0.0000i
1	0.7778 − 3.0000i	1.0458 + 2.2762i	2.5547 + 0.0321i
2	0.7386 − 1.5940i	0.8836 + 1.3060i	2.1840 + 0.0291i
3	0.9543 − 1.0935i	0.9705 + 1.0222i	2.0141 + 0.0066i
4	1.0001 − 1.0023i	0.9996 + 0.9998i	2.0000 + 0.0000i
5	1.0000 − 1.0000i	1.0000 + 1.0000i	2.0000 − 0.0000i

From the results in Table 3.3, we find an interesting phenomenon

$$z_1^k + z_2^k + z_3^k \approx 4, \qquad \text{for all } k \geq 1.$$

In fact, the following remarkable property is known.

Theorem 3.3 *(Dochev)* *For any starting point $z_1^0, \ldots, z_2^0$, the Durand-Kerner sequences satisfy*

$$\sum_{i=1}^{n} z_i^k = -a_1, \qquad k \geq 1.$$

Hence, the center of gravity for $\{z_i^k\}_{i=1}^n$ is equal to that of $\{z_i^*\}_{i=1}^n$.

Proof: (Dochev stated this result without proof). Let $z_1, z_2, \ldots, z_n$ be n distinct numbers and put

$$q(z) = p(z) - \prod_{i=1}^{n}(z - z_i).$$

Then $q(z)$ is a polynomial of degree $\leq n - 1$ and $q(z_j) = p(z_j), j = 1, 2, \ldots, n$. Hence, Lagrange's interpolation formula (cf. Section 5.2) yields

$$q(z) = \sum_{i=1}^{n} p(z_i)\ell_i(z).$$

Comparing the coefficient of z^{n-1} in both sides, we have

$$a_1 + \sum_{i=1}^{n} z_i = \sum_{i=1}^{n} \frac{p(z_i)}{\prod_{\substack{j=1 \\ j \neq i}}^{n}(z_i - z_j)}.$$

This implies

$$\sum_{i=1}^{n} \left\{ z_i - \frac{p(z_i)}{\prod_{\substack{j=1 \\ j \neq i}}^{n}(z_i - z_j)} \right\} = -a_1,$$

for any $\{z_i\}_{i=1}^{n}$ with $z_i \neq z_j (i \neq j)$, which proves Theorem 3.3. ∎

Exercises

(3.1) Show that from any $x_0 \in [1.2, 2.2]$ the Newton method converges to the solution of $f(x) = x^2 - 3 = 0$.

(3.2) Solve $f(x) = x^2 - 3 = 0$ by the Newton method starting from $x_0 = 1$. Write down $x_1, x_2, \ldots, x_{10}$.

(3.3) Solve $f(x) = x^2 - 3$ by the secant method starting from $x_0 = 1, x_1 = 0$. Write down $x_2, x_3, \ldots, x_{10}$.

(3.4) Set up a Newton iteration for computing the cube root of a given number c and apply it to $c = 3$ with $x_0 = 1$.

(3.5) Give an interval $[a, b]$ such that for any starting point $x_0 \in [a, b]$ the Newton method for computing $\sqrt[3]{3}$ converges.

(3.6) Let a be a positive number and $f(x) = a - \frac{1}{x}$. Show that from any starting point $x_0 \in [\frac{3}{4a}, \frac{5}{4a}]$, the Newton method converges to the solution of $f(x) = 0$. For $a = 6$, apply the Newton method with $x_0 = \frac{1}{8}$ to find an approximate value of $\frac{1}{6}$. Write down x_1, x_2.

(3.7) Apply the Newton method with $(x_0, y_0) = (0, 0)$ to solve the system
$$2x + x^2 - y - 2 = 0$$
$$-x + 4y + y^2 - 4 = 0$$

(3.8) Apply the Weierstrass method to find the zeros of the polynomial
$$p(z) = z^3 - 6z^2 + 11z - 6.$$

Chapter 4.

The Matrix Eigenvalue Problem

4.1 Eigenvalues and eigenvectors

Let A be an $n \times n$ matrix. A real or complex number λ is called an **eigenvalue** of the matrix A if there exists a vector $x \in C^n$, $x \neq 0$ such that

$$Ax = \lambda x. \tag{4.1}$$

The vector $x \neq 0$ is then called an **eigenvector** of A corresponding to the eigenvalue λ. The system (4.1) can be written as

$$(A - \lambda I)x = 0 \tag{4.2}$$

where I is the $n \times n$ identity matrix. Hence, the following statements are equivalent.

(i) λ is an eigenvalue of A.
(ii) The homogeneous system (4.2) has a nontrivial solution x.
(iii) The characteristic determinant $\det(A - \lambda I)$ is 0.

Expanding the characteristic determinant, we obtain a polynomial of degree n

$$\begin{aligned} p(\lambda) &= \det(A - \lambda I) \\ &= (-1)^n \lambda^n + (-1)^{n-1}(a_{11} + \ldots + a_{nn})\lambda^{n-1} + \ldots + \det A, \end{aligned}$$

---

matrix eigenvalue problem [行列の固有値問題]　　eigenvalue [固有値]　　eigenvector [固有ベクトル]　　characteristic determinant [特性行列式]

which is called the characteristic polynomial of A. The equation $\det(A - \lambda I) = 0$ is called the characteristic equation. Hence A has just n eigenvalues counting multiplicity. If A is real, so are the coefficients of the characteristic polynomial. Then the roots (the eigenvalues of A) are real or complex conjugates pairs.

If A is real symmetric, that is, $A = A^T$, then all eigenvalues of A are real.

We denote the eigenvalues of A by

$$\lambda_1, \lambda_2, \ldots, \lambda_n.$$

The sum of these n eigenvalues equals to the sum of the elements on the main diagonal of A, called the **trace** of A, and denoted by $\mathrm{Tr}(A)$. Thus

$$\mathrm{Tr}(A) \equiv \sum_{i=1}^{n} a_{ii} = \sum_{i=1}^{n} \lambda_i.$$

Moreover the determinant of A is equal to the product of its eigenvalues:

$$\det(A) = \prod_{i=1}^{n} \lambda_i.$$

Similar matrices An $n \times n$ matrix B is said to be similar to A if there is a nonsingular matrix T such that

$$B = T^{-1}AT.$$

The matrices A and B have the same eigenvalues, since

$$\det(B - \lambda I) = \det(T^{-1}(A - \lambda I)T) = \det(A - \lambda I).$$

Moreover, if x is an eigenvector of A, then $y = T^{-1}x$ is an eigenvector of B corresponding to the same eigenvalue.

Orthogonal matrices An $n \times n$ matrix A is said to be an orthogonal matrix if $AA^T = A^T A = I$. This implies that A is nonsingular and $A^{-1} = A^T$. The eigenvalues of an orthogonal matrix have absolute value 1.

Spectral shift If A has the eigenvalues $\lambda_1, \ldots, \lambda_n$, then $A - \alpha I$ has the eigenvalues

$$\lambda_1 - \alpha, \lambda_2 - \alpha, \ldots \lambda_n - \alpha.$$

---

characteristic polynomial [特性多項式]　characteristic equation [特性方程式]　conjugate [共役]　trace [トレース、跡]　determinant [行列式]　symmetric [対称]　similar matrix [相似行列]　orthogonal matrix [直交行列]　spectral shift [固有値のシフト]

4.2 Inclusion of eigenvalues

By inclusion we mean the determination of a region which includes the eigenvalues.

Theorem 4.1 *(Gerschgorin's theorem)*   *Let R_i denote the closed disk in the complex plane with center a_{ii} and radius*

$$r_i = \sum_{\substack{j=1 \\ j \neq i}}^{n} |a_{ij}|,$$

that is, $R_i = \{z \in C \mid |z - a_{ii}| \leq r_i\}$. Then we have the following statements.

(i) *All eigenvalues of A are in*

$$\hat{R} = \bigcup_{i=1}^{n} R_i.$$

(ii) *If $R_{i_1}, R_{i_2}, ..., R_{i_m}$ form a set S that is disjoint from the $n-m$ other R_i, then S contains precisely m eigenvalues of A.*

Proof: (i) Let λ be an eigenvalue of A and let x be an eigenvector corresponding to λ. Then $x \neq 0$, and

$$Ax = \lambda x, \qquad \|x\|_\infty > 0.$$

Suppose $\|x\|_\infty = |x_k|$. Considering the kth component of $Ax = \lambda x$, we get

$$(a_{kk} - \lambda)x_k = -\sum_{j \neq k} a_{kj} x_j$$

and

$$|a_{kk} - \lambda||x_k| = |(a_{kk} - \lambda)x_k| = \left|\sum_{\substack{j=1 \\ j \neq k}}^{n} a_{kj} x_j\right| \leq \sum_{\substack{j=1 \\ j \neq k}}^{n} |a_{kj}||x_j|$$

$$\leq \sum_{\substack{j=1 \\ j \neq k}}^{n} |a_{kj}||x_k| = r_k |x_k|,$$

---

inclusion of eigenvalues [固有値の包含]   closed disk [閉円板]   complex plane [複素平面]

that is, $|a_{kk} - \lambda| \leq r_k$. Hence $\lambda \in R_k \subseteq \hat{R}$.

(ii) We split the matrix A into its diagonal part D and off-diagonal part B to give $A = D + B$. Consider the matrix

$$A_\epsilon = D + \epsilon B.$$

We have that $A_\epsilon = D$ for $\epsilon = 0$ and $A_\epsilon = A$ for $\epsilon = 1$.

As ϵ varies from 0 to 1, any eigenvalue of A_ϵ is on a continuous path starting from a_{ii} with some i to an eigenvalue of A. Without loss of generality, we may assume $i_1 = 1, i_2 = 2, \ldots, i_m = m$ and let

$$R_i(\epsilon) = \{z \in C \,|\, |z - a_{ii}| \leq \epsilon r_i\}.$$

Then

$$S(\epsilon) = \bigcup_{i=1}^{m} R_i(\epsilon)$$

is disjoint from

$$\hat{S}(\epsilon) = \bigcup_{i=m+1}^{n} R_i(\epsilon).$$

By part (i) of this theorem and continuity of eigenvalues, any eigenvalue in $S(\epsilon)$ can not jump from $S(\epsilon)$ to $\hat{S}(\epsilon)$ as ϵ changes from 0 to 1. This implies $S(\epsilon)$ contains precisely m eigenvalues of A_ϵ since $S(0)$ contains m eigenvalues $a_{11}, \ldots, a_{mm}$ of $A_0 = D$ counting multiplicity. ∎

The Gerschgorin theorem is an important "inclusion theorem", which gives a region consisting of closed circular disks in the complex plane and including all the eigenvalues of a given matrix. Indeed, for each $i = 1, \ldots, n$, Gerschgorin's theorem determines a closed circular disk in the complex λ-plane with center a_{ii} and radius r_i. The result (i) of Theorem 4.1 states that each of the eigenvalues of A lies in one of these n disks. The result (ii) of Theorem 4.1 states that if m disks form a set S which is disjoint from the $n - m$ other disks, then S contains precisely m eigenvalues.

Example 4.1 Apply the Gerschgorin theorem to estimate the location of the eigenvalues of the matrix

$$A = \begin{bmatrix} 5 & 0 & 1 \\ -0.2 & -2 & -1 \\ 0.5 & -1 & 3 \end{bmatrix}.$$

---

off-diagonal part [非対角部分]   inclusion theorem [包含定理]

4.3 The power method

Solution The Gerschgorin disks for the matrix A are as follows.

R_1 :   center 5,   radius 1
R_2 :   center -2,   radius 1.2
R_3 :   center 3,   radius 1.5

R_2 contains one eigenvalue of A and $R_1 \cup R_3$ contains two eigenvalues of A as shown by Figure 4.1. $\diamond$

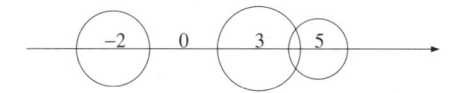

Figure 4.1: Gerschgorin's disks

Note: The matrix in Example 4.1 has eigenvalues $\lambda_1 \approx 5.2498, \lambda_2 \approx -2.1899$, and $\lambda_3 \approx 2.9400$.

4.3 The power method

The power method is a simple procedure for finding an eigenvalue of an $n \times n$ matrix A which is larger in modulus than any other eigenvalue. In this method we start from an n dimensional vector $x^0 \neq 0$ and compute

$$x^{k+1} = Ax^k, \quad \text{for } k \geq 0. \tag{4.3}$$

The method is based upon the following theorem.

Theorem 4.2 *Suppose that the $n \times n$ matrix A has n eigenvalues $\lambda_1, \ldots, \lambda_n$ with an associated collection of linearly independent eigenvectors $\{v^1, v^2, \ldots, v^n\}$. Moreover, we assume that λ_1 is real and*

$$|\lambda_1| > |\lambda_i|, \quad i = 2, 3, \ldots n.$$

If $\{x^0, v^2, \ldots, v^n\}$ is linearly independent, then there exists an index j with $1 \leq j \leq n$ such that the sequnce $\{x^k\}$ generated by (4.3) from x^0 satisfies

$$\lim_{k \to \infty} \frac{x_j^{k+1}}{x_j^k} = \lambda_1.$$

---

power method [累乗法]   linearly independent [線形独立、一次独立]

Proof: Since $\{v^1, v^2, \ldots, v^n\}$ is linearly independent, there are constants $\alpha_1, \ldots, \alpha_n$ such that

$$x^0 = \sum_{i=1}^{n} \alpha_i v^i.$$

From the assumption that $\{x^0, v^2, \ldots, v^n\}$ is linearly independent, we have $\alpha_1 \neq 0$. Multiplying both sides of this equation by A^k, we obtain

$$x^k = A^k x^0 = \sum_{i=1}^{n} A^k \alpha_i v^i = \sum_{i=1}^{n} \alpha_i \lambda_i^k v^i = \lambda_1^k \sum_{i=1}^{n} \alpha_i \left(\frac{\lambda_i}{\lambda_1}\right)^k v^i,$$

and similarly

$$x^{k-1} = \lambda_1^{k-1} \sum_{i=1}^{n} \alpha_i \left(\frac{\lambda_i}{\lambda_1}\right)^{k-1} v^i.$$

Since $|\lambda_1| > |\lambda_i|$ for all $i = 2, 3, \ldots n$, we have

$$\lim_{k \to \infty} \left(\frac{\lambda_i}{\lambda_1}\right)^k = 0, \qquad \text{for } i = 2, \ldots, n.$$

Hence, if v_j^1 is a non-zero component of v^1, then for all sufficiently large k

$$x_j^k = \lambda_1^k \left[\alpha_1 v_j^1 + \sum_{i=2}^{n} \alpha_i \left(\frac{\lambda_i}{\lambda_1}\right)^k v_j^i\right] \neq 0.$$

Therefore

$$\frac{x_j^{k+1}}{x_j^k} = \lambda_1 \frac{\alpha_1 v_j^1 + \sum_{i=2}^{n} \alpha_i \left(\frac{\lambda_i}{\lambda_1}\right)^{k+1} v_j^i}{\alpha_1 v_j^1 + \sum_{i=2}^{n} \alpha_i \left(\frac{\lambda_i}{\lambda_1}\right)^k v_j^i} \to \lambda_1, \qquad \text{as } k \to \infty.$$

∎

In actual computation, the condition that the set of vectors $\{x^0, v^2, \ldots, v^n\}$ is linearly independent is satisfied except in the rare case that x^0 is exactly a linear combination of $v^2, \ldots, v^n$.

Example 4.2 Apply the power method (3 steps) to the matrix in Example 4.1 starting from $x^0 = [1, 1, 1]^T$.

Solution Starting from x^0, we compute

$$x^1 = Ax^0 = \begin{bmatrix} 6 \\ -3.2 \\ 2.5 \end{bmatrix}, x^2 = Ax^1 = \begin{bmatrix} 32.5 \\ 2.7 \\ 13.7 \end{bmatrix}, x^3 = Ax^2 = \begin{bmatrix} 176.2 \\ -25.6 \\ 54.65 \end{bmatrix}$$

and

$$\frac{x_1^1}{x_1^0} = 6, \quad \frac{x_1^2}{x_1^1} = 5.4167, \quad \frac{x_1^3}{x_1^2} = 5.4215. \qquad \diamondsuit$$

In practice, we normalize x^k to prevent the overflow of Ax^k as k increases. For instance, we use the following procedure.

$$\begin{aligned} y^{k+1} &= Ax^k \\ x^{k+1} &= \frac{y^{k+1}}{\|y^{k+1}\|_\infty}. \end{aligned}$$

Frequently, after we computed an approximate eigenvalue $\hat{\lambda}_1$, we want to obtain other eigenvalues of A. In such case, we can replace A by $A - \hat{\lambda}_1 I$ in the power method and compute an eigenvalue $\lambda_j - \hat{\lambda}_1$ of $A - \hat{\lambda}_1 I$, which satisfies

$$|\lambda_j - \hat{\lambda}_1| > |\lambda_i - \hat{\lambda}_1|, \qquad i = 1, 2, \ldots, n, \ i \neq j.$$

This procedure is known as a deflation technique.

4.4 Householder tridiagonalization

In Sections 4.4 and 4.5, we study a method for computing all the eigenvalues of a real symmetric matrix A. In the first stage, we apply Householder's method which reduces the given matrix to a **tridiagonal** matrix. In the second stage, the tridiagonal matrix is factorized in the QR form, where Q is orthogonal and R is upper triangular. This method is widely used in practice.

Let $u \in R^n$ be a vector with $\|u\|_2 = 1$. The matrix

$$P = I - 2uu^T$$

is called the Householder matrix.

---

deflation technique [デフレーション（抜取り）技法]　tridiagonalization [3 重対角化] ($|i-j| \geq 2$ のとき $a_{i,j} = 0$ なる行列 $A = (a_{i,j})$ を 3 重対角行列という)

Proposition 4.1 *The Householder matrix has the following properties.*
1. $P = P^T = P^{-1}$.
2. $\|Px\|_2 = \|x\|_2$, for all $x \in R^n$.
3. For any $x, y \in R^n$, if $x \neq y$ and $\|y\|_2 = \|x\|_2$, then $Px = y$ for

$$u = \frac{x-y}{\|x-y\|_2}.$$

Proof: 1. By straightforward calculation, we obtain

$$P = I - 2uu^T = I - 2(uu^T)^T = (I - 2uu^T)^T = P^T,$$

and

$$\begin{aligned}PP &= (I - 2uu^T)(I - 2uu^T) \\ &= I + 4uu^Tuu^T - 4uu^T \\ &= I + 4uu^T - 4uu^T \\ &= I,\end{aligned}$$

where we have used $u^T u = \|u\|_2^2 = 1$.
2. Using $P^T P = P^{-1} P = I$, we get

$$\|Px\|_2^2 = (Px)^T Px = x^T P^T Px = x^T x = \|x\|_2^2.$$

3. Since $\|y\|_2 = \|x\|_2$, we have

$$\|x-y\|_2^2 = x^T x - 2x^T y + y^T y = 2(x^T x - x^T y) = 2x^T(x-y).$$

Hence

$$\begin{aligned}Px &= x - 2\frac{(x-y)(x-y)^T}{\|x-y\|_2^2} x \\ &= x - 2\frac{(x-y)^T x}{\|x-y\|_2^2}(x-y) \\ &= x - (x-y) \\ &= y.\end{aligned}$$

∎

Using $n-2$ Householder matrices $P_1, P_2, \ldots, P_{n-2}$, the Householder method reduces an $n \times n$ real symmetric matrix A by $n-2$ successive

4.4 Householder tridiagonalization

similarity transformations to tridiagonal form. The $n-2$ similarity transformations that produce the matrices $A_1 = [a_{ij}^{(1)}]$, $A_2 = [a_{ij}^{(2)}]$, etc., from the given matrix $A_0 = A = [a_{ij}^{(0)}]$ successively can be done as follows.

$$\begin{aligned} A_1 &= P_1 A_0 P_1 \\ A_2 &= P_2 A_1 P_2 \\ &\cdots\cdots \\ A_{n-2} &= P_{n-2} A_{n-1} P_{n-2} \\ &= P_{n-2} \ldots P_1 A P_1 \ldots P_{n-2} \\ &= (P_1 \ldots P_{n-2})^{-1} A (P_1 \ldots P_{n-2}). \end{aligned}$$

In the first step, P_1 eliminates $a_{31}^{(0)}, \ldots, a_{n1}^{(0)}$ in column 1 and $a_{13}^{(0)}, \ldots, a_{1n}^{(0)}$ in row 1. In the second step, P_2 eliminates $a_{42}^{(1)}, \ldots, a_{n2}^{(1)}$ in column 2 and $a_{24}^{(1)}, \ldots, a_{2n}^{(1)}$ in row 2, and so on. We illustrate these transformations for a 5×5 matrix:

$$\begin{bmatrix} * & * & & & \\ * & * & * & * & * \\ & * & * & * & * \\ & * & * & * & * \\ & * & * & * & * \end{bmatrix} \begin{bmatrix} * & * & & & \\ * & * & * & & \\ & * & * & * & * \\ & & * & * & * \\ & & * & * & * \end{bmatrix} \begin{bmatrix} * & * & & & \\ * & * & * & & \\ & * & * & * & \\ & & * & * & * \\ & & & * & * \end{bmatrix}.$$

Determination of $P_1, \ldots, P_{n-2}$ Let $a_i^{(i-1)}$ be the ith column of A_{i-1}. We then define the ith column of A_i by

$$a_{ji}^{(i)} = a_{ji}^{(i-1)}, \quad j = 1, 2, \ldots, i$$

$$a_{i+1\,i}^{(i)} = -\operatorname{sgn}(a_{i+1\,i}^{(i-1)}) S_i,$$

and

$$a_{ji}^{(i)} = 0, \quad j = i+2, \ldots, n,$$

where

$$S_i = \sqrt{(a_{i+1\,i}^{(i-1)})^2 + \ldots + (a_{ni}^{(i-1)})^2}.$$

The use of $-\operatorname{sgn}(a_{i+1\,i}^{(i-1)})$ is to avoid loss of significant digits in computing $\|a_i^{(i-1)} - a_i^{(i)}\|$.

All these P_i are of the form

$$P_i = I - 2 u^{(i)} (u^{(i)})^T, \quad i = 1, 2, \ldots, n-2,$$

where
$$u^{(i)} = \frac{a_i^{(i-1)} - a_i^{(i)}}{\|a_i^{(i-1)} - a_i^{(i)}\|}.$$

Now it is easy to find the components of $u^{(i)} = [u_1^{(i)}, \ldots, u_n^{(i)}]^T$;

$$u_j^{(i)} = 0, \quad j = 1, 2, \ldots, i$$

$$u_{i+1}^{(i)} = \mathrm{sgn}(a_{i+1\,i}^{(i-1)}) \sqrt{\frac{1}{2}\left(1 + \frac{|a_{i+1\,i}^{(i-1)}|}{S_i}\right)}$$

$$u_j^{(i)} = \frac{a_{ji}^{(i-1)}}{2|u_{i+1}^{(i)}|S_i}, \quad j = i+2, \ldots, n.$$

Example 4.3 Apply Householder's method to tridiagonalize the symmetric matrix
$$A = \begin{bmatrix} 2 & 3 & -4 \\ 3 & 2 & 0 \\ -4 & 0 & 1 \end{bmatrix}.$$

Solution Only one Householder transformation P_1 is needed to tridiagonalize the 3×3 matrix. Let $A_0 = A = (a_{ij}^{(0)})$. Since $a_{21}^{(0)} = 3 > 0$, $\mathrm{sgn}(a_{21}^{(0)}) = 1$.

$$S_1 = \sqrt{(a_{21}^{(0)})^2 + (a_{31}^{(0)})^2} = \sqrt{3^2 + (-4)^2} = 5$$

$$a_1^{(0)} = \begin{bmatrix} 2 \\ 3 \\ -4 \end{bmatrix}, \quad a_1^{(1)} = \begin{bmatrix} 2 \\ -\mathrm{sgn}(a_{21}^{(0)})S_1 \\ 0 \end{bmatrix} = \begin{bmatrix} 2 \\ -5 \\ 0 \end{bmatrix}$$

$$u^{(1)} = \frac{a_1^{(0)} - a_1^{(1)}}{\|a_1^{(0)} - a_1^{(1)}\|} = \frac{1}{\sqrt{64+16}} \begin{bmatrix} 0 \\ 8 \\ -4 \end{bmatrix} = \frac{1}{\sqrt{5}} \begin{bmatrix} 0 \\ 2 \\ -1 \end{bmatrix}$$

$$u^{(1)}(u^{(1)})^T = \begin{bmatrix} 0 \\ 2/\sqrt{5} \\ -1/\sqrt{5} \end{bmatrix} \begin{bmatrix} 0 & 2/\sqrt{5} & -1/\sqrt{5} \end{bmatrix} = \begin{bmatrix} 0 & 0 & 0 \\ 0 & 4/5 & -2/5 \\ 0 & -2/5 & 1/5 \end{bmatrix}$$

$$P_1 = I - 2u^{(1)}(u^{(1)})^T = \begin{bmatrix} 1 & 0 & 0 \\ 0 & -3/5 & 4/5 \\ 0 & 4/5 & 3/5 \end{bmatrix}$$

$$P_1AP_1 = \begin{bmatrix} 2 & 3 & -4 \\ -5 & -6/5 & 4/5 \\ 0 & 8/5 & 3/5 \end{bmatrix} P_1 = \begin{bmatrix} 2 & -5 & 0 \\ -5 & 34/25 & -12/25 \\ 0 & -12/25 & 41/25 \end{bmatrix}.$$

$\diamond$

4.5 The QR-factorization method

A QR-factorization of a given square matrix A is of the form

$$A = QR$$

where Q is orthogonal and R is upper triangular. We study the QR-factorization method for a real symmetric tridiagonal matrix. It can be proved that every real symmetric matrix has an QR factorization. In this method, we compute $A_1, A_2, \ldots$ according to the following rule.

$$A_1 = A = Q_1 R_1$$

where Q_1 is orthogonal and R_1 is upper triangular. Then compute

$$A_2 = R_1 Q_1$$

and factor
$$A_2 = Q_2 R_2.$$

General step:

    Factor $A_k = Q_k R_k$
    Compute $A_{k+1} = R_k Q_k.$

We have $R_k = Q_k^{-1} A_k$ and thus

$$\begin{aligned} A_{k+1} &= Q_k^{-1} A_k Q_k = Q_k^{-1} Q_{k-1}^{-1} A_{k-1} Q_{k-1} Q_k = \ldots \\ &= Q_k^{-1} \ldots Q_1^{-1} A Q_1 \ldots Q_k = (Q_1 \ldots Q_k)^{-1} A (Q_1 \ldots Q_k). \end{aligned}$$

Hence A_{k+1} and A are similar, and thus they have the same eigenvalues. The following result is known [Kreyszig 1993].

Theorem 4.3 *If the eigenvalues of a symmetric matrix A are different in absolute value, say, if we have*

$$|\lambda_1| > |\lambda_2| > \ldots > |\lambda_n| > 0,$$

then
$$\lim_{k \to \infty} A_k = D$$

where D is diagonal with diagonal elements $d_{ii} = \lambda_i$, $i = 1, 2, \ldots, n$.

  QR-factorization [QR 分解] tridiagonal matrix [三重対角行列]

QR-factorization of a tridiagonal matrix   Let A be an $n \times n$ tridiagonal matrix obtained from a real symmetric matrix by Householder's method

$$A = \begin{bmatrix} a_1 & b_2 & & & \\ b_2 & a_2 & b_3 & & \\ & \ddots & \ddots & \ddots & \\ & & \ddots & \ddots & b_n \\ & & & b_n & a_n \end{bmatrix}.$$

The matrix A has $n-1$ generally nonzero entries $b_2, \ldots b_n$ below the main diagonal. We multiply A from the left by an orthogonal matrix C_1 such that $C_1 A = [a_{ij}^{(1)}]$ has $b_2^{(1)} \equiv a_{21}^{(1)} = 0$. We multiply this by an orthogonal matrix C_2 such that $C_2 C_1 A = [a_{ij}^{(2)}]$ has $b_3^{(2)} \equiv a_{32}^{(2)} = 0$, etc. After $n-1$ such multiplications, we are left with an upper triangular matrix R_0, namely

$$C_{n-1} C_{n-2} \ldots C_1 A = R_0.$$

The construction of these matrices C_k is very simple. Let $A_0 = A$ ($a_i^{(0)} = a_i, b_i^{(0)} = b_i$). For $k = 1, 2, \ldots, n-1$, $C_k = (c_{ij}^{(k)})$ is defined by

$$v_k = \frac{b_{k+1}^{(k-1)}}{a_k^{(k-1)}}$$

$$\begin{aligned} c_{ij}^{(k)} &= 0, \quad i \neq j & &\text{except for} \quad c_{k+1,k}^{(k)}, \quad c_{k,k+1}^{(k)} \\ c_{ii}^{(k)} &= 1, \quad i = 1, \ldots n & &\text{except for} \quad c_{k,k}^{(k)}, \quad c_{k+1,k+1}^{(k)} \\ c_{kk}^{(k)} &= \frac{1}{\sqrt{1+v_k^2}}, & c_{k,k+1}^{(k)} &= \frac{v_k}{\sqrt{1+v_k^2}} \\ c_{k+1,k}^{(k)} &= -c_{k,k+1}^{(k)}, & c_{k+1,k+1}^{(k)} &= c_{kk}^{(k)}. \end{aligned}$$

Example 4.4   Give a QR factorization of the matrix

$$A = \begin{bmatrix} 3 & 1 & 0 \\ 1 & 3 & 1 \\ 0 & 1 & 3 \end{bmatrix}$$

Solution   We will use two orthogonal matrices C_1 and C_2 to give a QR factorization of the matrix A.

$$v_1 = \frac{b_2}{a_1} = \frac{1}{3}$$

$$C_1 = \begin{bmatrix} 3/\sqrt{10} & 1/\sqrt{10} & 0 \\ -1/\sqrt{10} & 3/\sqrt{10} & 0 \\ 0 & 0 & 1 \end{bmatrix}$$

$$C_1 A = \begin{bmatrix} \sqrt{10} & 6/\sqrt{10} & 1/\sqrt{10} \\ 0 & 8/\sqrt{10} & 3/\sqrt{10} \\ 0 & 1 & 3 \end{bmatrix}$$

$$v_2 = \frac{b_3^{(1)}}{a_{22}^{(1)}} = \frac{\sqrt{10}}{8}$$

$$C_2 = \begin{bmatrix} 1 & 0 & 0 \\ 0 & 8/\sqrt{74} & \sqrt{10/74} \\ 0 & -\sqrt{10/74} & 8/\sqrt{74} \end{bmatrix}$$

$$C_2 C_1 A = \begin{bmatrix} \sqrt{10} & 6/\sqrt{10} & 1/\sqrt{10} \\ 0 & \sqrt{74/10} & 54/\sqrt{740} \\ 0 & 0 & 21/\sqrt{74} \end{bmatrix}$$

Let $Q = C_1^T C_2^T$ and $R = C_2 C_1 A$. Then Q is orthogonal and $R = Q^T A$. We thus obtain $A = QR$, which is a QR factorization of A. $\diamondsuit$

Remark The QR method is also applicable to not necessarily symmetric matrices. See [Yamamoto 2003] for its convergence.

Exercises

(4.1) Apply the Gerschgorin theorem to determine the interval containing eigenvalues of the symmetric matrix

$$A = \begin{bmatrix} 1 & 1 & -1 \\ 1 & -1 & -2 \\ -1 & -2 & 8 \end{bmatrix}.$$

(4.2) Use the Gerschgorin theorem to show that if λ is the minimal eigenvalue of the matrix

$$A = \begin{bmatrix} 2 & -1 & 0 \\ -1 & 2 & -1 \\ 0 & -1 & 2 \end{bmatrix}$$

then $|\lambda - 4| = \rho(A - 4I)$, where ρ denotes the spectral radius.

(4.3) Apply the power method to find the eigenvalue with maximum absolute value of the matrix

$$A = \begin{bmatrix} 2 & -1 & 0 \\ -1 & 2 & -1 \\ 0 & -1 & 2 \end{bmatrix}.$$

(4.4) Let $x = [2, 2, 1]^T$ and $y = [3, 0, 0]^T$. Find a symmetric and orthogonal matrix $P \in R^{3 \times 3}$ such that $Px = y$.

(4.5) Compute the QR factorization of the matrix

$$A = \begin{bmatrix} 4 & 0 & 0 \\ 0 & 3 & 1 \\ 0 & 1 & 2 \end{bmatrix}.$$

Chapter 5.

Interpolation Polynomials

5.1 Introduction

For given $n+1$ distinct real numbers

$$x_0 < x_1 < \ldots < x_n$$

and $n+1$ function values at these points

$$f_0, \ f_1, \ \ldots, \ f_n,$$

interpolation is to find a function p such that

$$p(x_i) = f_i, \qquad i = 0, 1, 2, \ldots, n. \tag{5.1}$$

Using the function p, we get approximate values of f for x in the interval (x_0, x_n) ("interpolation") or sometimes outside of the interval $[x_0, x_n]$ ("extrapolation").

If the function p is a polynomial, then the interpolation is called a polynomial interpolation and p is called an interpolation polynomial. Moreover, if the function p_n is a polynomial of degree $\leq n$ and satisfies (5.1), then the polynomial p_n is uniquely determined as is shown below. We say p_n is the nth **interpolation polynomial** of f. These points $x_0, x_1, \ldots, x_n$ are called **nodes**.

---

interpolation [補間]　　interpolation polynomial [補間多項式]　　extrapolation [外挿]　　polynomial interpolation [多項式補間]　　node [分点]

The nth degree polynomial for real variables and with real coefficients is of the form
$$p_n(x) = a_0 + a_1 x + \ldots + a_n x^n,$$
where a_i, $i = 0, 1, \ldots, n$ are real numbers. For this to be an interpolation polynomial, one has to find these real coefficients $a_i, i = 0, 1, \ldots, n$ such that
$$p_n(x_i) = a_0 + a_1 x_i + \ldots + a_n x_i^n = f_i, \qquad i = 0, 1, \ldots, n. \qquad (5.2)$$
Let
$$V = \begin{bmatrix} 1 & x_0 & \ldots & x_0^n \\ 1 & x_1 & \ldots & x_1^n \\ \vdots & \vdots & & \vdots \\ 1 & x_n & \ldots & x_n^n \end{bmatrix}, \quad a = \begin{bmatrix} a_0 \\ a_1 \\ \vdots \\ a_n \end{bmatrix} \quad \text{and} \quad b = \begin{bmatrix} f_0 \\ f_1 \\ \vdots \\ f_n \end{bmatrix}.$$
Then the system of (5.2) can be written as
$$Va = b. \qquad (5.3)$$
This is a system of $n+1$ linear equations with $n+1$ unknowns. The matrix V is called a Vandermonde matrix. It is well known that
$$\det(V) = (-1)^{\frac{n(n+1)}{2}} \prod_{0 \leq i < j \leq n} (x_i - x_j).$$
Since the points x_i are distinct, V is nonsingular and the system (5.3) has a unique solution. This shows that there exists a unique polynomial p_n of degree $\leq n$ satisfying (5.1). Observe that a_n may be zero so that degree of p_n may be smaller than n.

We will study several methods to define p_n. For given data, these methods give the same polynomial, but in different forms, which differ in the amount of computation.

5.2 The Lagrange interpolation formula

The Lagrange interpolation is defined by using $n+1$ special polynomials $\ell_i(x_i)$ $(i = 0, 1, \ldots, n)$ of degree n. These polynomials have the form
$$\ell_i(x) = \frac{(x - x_0) \ldots (x - x_{i-1})(x - x_{i+1}) \ldots (x - x_n)}{(x_i - x_0) \ldots (x_i - x_{i-1})(x_i - x_{i+1}) \ldots (x_i - x_n)}, \quad i = 0, \ldots, n.$$
interpolation formula [補間公式]

5.2 The Lagrange interpolation formula

It is easy to see that they have the property

$$\ell_i(x_i) = 1, \quad \ell_i(x_j) = 0, \quad \text{for } j \neq i, \tag{5.4}$$

that is, 1 at x_i and 0 at other nodes. The polynomials $\ell_i(x)$ are called the Lagrange factors or the fundamental polynomial for point-wise interpolation.

Multiplying each f_i ($0 \leq i \leq n$) by $\ell_i(x)$, and then adding up the resulting polynomials, we get the **Lagrange interpolation formula**

$$p_n(x) = \sum_{i=0}^{n} \ell_i(x) f_i,$$

which satisfies $p_n(x_i) = f_i, i = 0, 1, \ldots, n$, that is, it passes through $n+1$ points $(x_i, f_i), i = 0, 1, \ldots, n$.

The Lagrange interpolation

$$p_1(x) = \ell_0(x) f_0 + \ell_1(x) f_1,$$

uses a line segment that passes through two points (x_0, f_0) and (x_1, f_1), where ℓ_0 and ℓ_1 are polynomials of degree 1 and satisfy

$$\ell_0(x_0) = 1, \qquad \ell_0(x_1) = 0$$
$$\ell_1(x_0) = 0, \qquad \ell_1(x_1) = 1.$$

Obviously

$$\ell_0(x) = \frac{x - x_1}{x_0 - x_1}, \qquad \ell_1(x) = \frac{x - x_0}{x_1 - x_0}.$$

Example 5.1 Use the Lagrange interpolation formula to construct a quadratic function p_2 which satisfies

$$p_2(0) = 2, \qquad p_2(1) = -1, \qquad p_2(2) = 4.$$

Solution Let

$$x_0 = 0, \qquad x_1 = 1, \qquad x_2 = 2$$
$$f_0 = 2, \qquad f_1 = -1, \qquad f_2 = 4.$$

Lagrange factor [ラグランジュ因子]　　fundamental polynomial [基本多項式]
point-wise [点毎、各点]

By the Lagrange interpolation formula, we have

$$p_2(x) = \ell_0(x)f_0 + \ell_1(x)f_1 + \ell_2(x)f_2 ,$$

where, as shown by Figure 5.1,

$$\ell_0(x) = \frac{(x-x_1)(x-x_2)}{(x_0-x_1)(x_0-x_2)} = \frac{(x-1)(x-2)}{2}$$

$$\ell_1(x) = \frac{(x-x_0)(x-x_2)}{(x_1-x_0)(x_1-x_2)} = \frac{x(x-2)}{-1}$$

$$\ell_2(x) = \frac{(x-x_0)(x-x_1)}{(x_2-x_0)(x_2-x_1)} = \frac{x(x-1)}{2}.$$

Thus

$$p_2(x) = (x-1)(x-2) + x(x-2) + 2x(x-1) = 4x^2 - 7x + 2. \quad \diamond$$

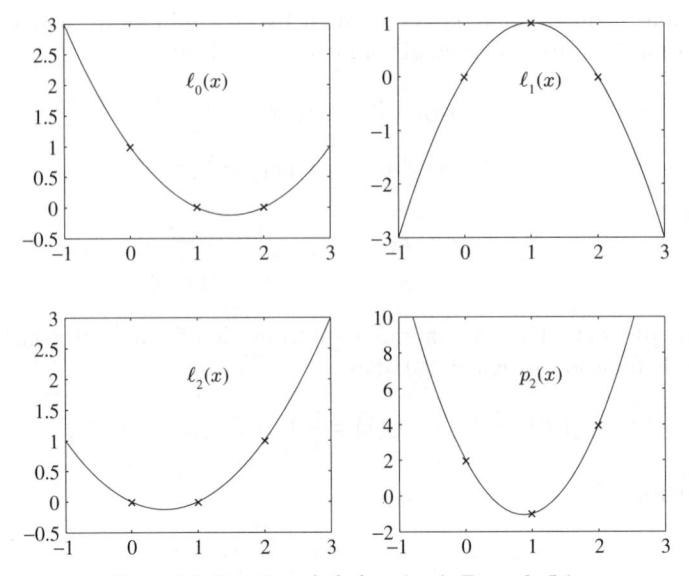

Figure 5.1: Functions ℓ_0, ℓ_1, ℓ_2 and p_2 in Example 5.1

5.3 The Newton divided difference interpolation formula

Suppose that p_n is the Lagrange interpolation polynomial of degree $\leq n$, which satisfies
$$p_n(x_i) = f_i, \qquad i = 0, 1, \ldots n,$$
at the distinct nodes $x_i, i = 0, 1, \ldots, n$. Now we want to add a new node x_{n+1} and define an interpolation polynomial p_{n+1} of degree $\leq n+1$ such that
$$p_{n+1}(x_i) = f_i, \qquad i = 0, 1, \ldots, n+1.$$

If we use the Lagrange interpolation formula, we need to construct $n+2$ new polynomials ℓ_i of degree $\leq n+1$, $i = 0, 1, 2, \ldots, n+1$. The work done in calculating p_n does not lessen the work needed to calculate p_{n+1}. With the Lagrange interpolation formula, it is inconvenient to pass from one interpolation to another of degree one greater.

The Newton divided difference interpolation formula uses the previous calculations and simply add another term to increase the degree of the interpolation polynomials.

Let
$$p_{n+1}(x) = p_n(x) + q_{n+1}(x),$$
where q_{n+1} is the correction term. Since $p_n(x_i) = p_{n+1}(x_i) = f_i$, $i = 0, 1, \ldots, n$, we have
$$q_{n+1}(x_i) = 0, \qquad i = 0, 1, \ldots, n.$$

Moreover, q_{n+1} is a polynomial of degree $\leq n+1$ because so is p_{n+1} whereas p_n is of degree $\leq n$. Hence q_{n+1} must be of the form
$$q_{n+1}(x) = a_{n+1}(x - x_0)(x - x_1) \cdots (x - x_n).$$

Comparing the coefficient a_{n+1} of x^{n+1} in
$$p_{n+1}(x) = p_n(x) + a_{n+1}(x - x_0)(x - x_1) \ldots (x - x_n) \tag{5.5}$$
with that of the Lagrange interpolation formula
$$p_{n+1}(x) = \sum_{i=0}^{n+1} \ell_i(x) f_i$$

---

correction term [修正項, 補正項]

we obtian

$$a_{n+1} = \sum_{i=0}^{n+1} \frac{f_i}{(x_i - x_0)\ldots(x_i - x_{i-1})(x_i - x_{i+1})\ldots(x_i - x_{n+1})}. \quad (5.6)$$

Furthermore, we have

$$\frac{1}{(x_i - x_0)\ldots(x_i - x_{i-1})(x_i - x_{i+1})\ldots(x_i - x_{n+1})}$$
$$= \frac{1}{x_{n+1} - x_0} \left[\frac{1}{(x_i - x_1)\ldots(x_i - x_{n+1})} - \frac{1}{(x_i - x_0)\ldots(x_i - x_n)} \right],$$
$$1 \leq i \leq n$$

and

$$\sum_{i=0}^{n+1} \frac{f_i}{(x_i - x_0)\ldots(x_i - x_{i-1})(x_i - x_{i+1})\ldots(x_i - x_{n+1})}$$
$$= \frac{f_0}{(x_0 - x_1)\ldots(x_0 - x_{n+1})} + \frac{f_{n+1}}{(x_{n+1} - x_0)\ldots(x_{n+1} - x_n)}$$
$$+ \sum_{i=1}^{n} \frac{f_i}{(x_i - x_0)\ldots(x_i - x_{i-1})(x_i - x_{i+1})\ldots(x_i - x_{n+1})}$$
$$= \frac{1}{x_{n+1} - x_0} \left[\frac{f_{n+1}}{(x_{n+1} - x_1)\ldots(x_{n+1} - x_n)} - \frac{f_0}{(x_0 - x_1)\ldots(x_0 - x_n)} \right]$$
$$+ \frac{1}{x_{n+1} - x_0} \sum_{i=1}^{n} \left[\frac{f_i}{(x_i - x_1)\ldots(x_i - x_{i-1})(x_i - x_{i+1})\ldots(x_i - x_{n+1})} \right.$$
$$\left. - \frac{f_i}{(x_i - x_0)\ldots(x_i - x_{i-1})(x_i - x_{i+1})\ldots(x_i - x_n)} \right]$$
$$= \frac{1}{x_{n+1} - x_0} \left[\sum_{i=1}^{n+1} \frac{f_i}{(x_i - x_1)\ldots(x_i - x_{i-1})(x_i - x_{i+1})\ldots(x_i - x_{n+1})} \right.$$
$$\left. - \sum_{i=0}^{n} \frac{f_i}{(x_i - x_0)\ldots(x_i - x_{i-1})(x_i - x_{i+1})\ldots(x_i - x_n)} \right].$$

Hence, if we denote the right-hand side of the expression (5.6) by $f[x_0, x_1, \ldots, x_{n+1}]$, then we have

$$f[x_0, x_1, \ldots, x_{n+1}] = \frac{f[x_1, x_2, \ldots, x_{n+1}] - f[x_0, x_1, \ldots, x_n]}{x_{n+1} - x_0},$$

5.3 The Newton divided difference interpolation formula

where, by definition,

$$f[x_1, x_2, \ldots, x_{n+1}] = \sum_{i=1}^{n+1} \frac{f_i}{(x_i - x_1)\ldots(x_i - x_{i-1})(x_i - x_{i+1})\ldots(x_i - x_{n+1})}$$

and

$$f[x_0, x_1, \ldots, x_n] = \sum_{i=0}^{n} \frac{f_i}{(x_i - x_0)\ldots(x_i - x_{i-1})(x_i - x_{i+1})\ldots(x_i - x_n)}.$$

Note that the expression $f[x_0, \ldots, x_n]$ is symmetric with respect to $x_0, x_1, \ldots, x_n$, that is,

$$f[x_0, x_1, \ldots, x_n] = f[x_{\sigma_0}, x_{\sigma_1}, \ldots, x_{\sigma_n}]$$

for any permutation $\{\sigma_0, \sigma_1, \ldots \sigma_n\}$ of $\{0, 1, 2, \ldots, n\}$.

$f[x_0, x_1, \ldots, x_n]$ is called the **nth-order divided difference** of f. Thus it follows from (5.5) that

$$p_{n+1}(x) = p_n(x) + (x - x_0)(x - x_1)\ldots(x - x_n)f[x_0, x_1, \ldots, x_{n+1}].$$

Observing $p_0(x) = f_0$ and repeating the argument above for $n = 0, 1, \cdots, n$, we obtain

$$\begin{aligned} p_0(x) &= f_0 \\ p_1(x) &= f_0 + (x - x_0)f[x_0, x_1] \\ p_2(x) &= f_0 + (x - x_0)f[x_0, x_1] + (x - x_0)(x - x_1)f[x_0, x_1, x_2] \\ &\cdots \\ p_n(x) &= f_0 + (x - x_0)f[x_0, x_1] + (x - x_0)(x - x_1)f[x_0, x_1, x_2] \\ &+ \cdots + (x - x_0)(x - x_1)\cdots(x - x_{n-1})f[x_0, x_1, \cdots, x_n]. \end{aligned}$$

This is called the **Newton divided difference interpolation formula.**

A standard idea in the Newton interpolation is to give a simple definition of coefficients a_i of the interpolation polynomial

$$p_n(x) = a_0 + a_1(x - x_0) + a_2(x - x_0)(x - x_1) + \ldots + a_n(x - x_0)\ldots(x - x_{n-1})$$

---

permutation [置換, 順列] nth–order divided difference [n 階差分商]

by using previous work, which can be done by the use of divided differences. From (5.6) and the definition of divided differences, we have

$$a_1 = f[x_0, x_1] = \frac{f_1 - f_0}{x_1 - x_0}$$

$$a_2 = f[x_0, x_1, x_2] = \frac{f[x_1, x_2] - f[x_0, x_1]}{x_2 - x_0}$$

$$\vdots$$

$$a_n = f[x_0, x_1, \ldots x_n] = \frac{f[x_1, \ldots, x_n] - f[x_0, \ldots, x_{n-1}]}{x_n - x_0}.$$

The reader should note that the coefficents a_n is computed by using the nodes $\{x_i\}_{i=0}^n$ and the function values $\{f_i\}_{i=0}^n$ at the nodes.
For $n = 0$,
$$p_0(x) \equiv a_0 = f_0.$$
For $n = 1$
$$p_1(x) = p_0(x) + a_1(x - x_0) = f_0 + a_1(x - x_0).$$

Since $f_1 = p_1(x_1)$, we have

$$a_1 = \frac{f_1 - f_0}{x_1 - x_0}.$$

For $n = 2$,
$$p_2(x) = p_1(x) + a_2(x - x_0)(x - x_1),$$
where

$$a_2 = \frac{f[x_1, x_2] - f[x_0, x_1]}{x_2 - x_0}$$

$$= \frac{1}{x_2 - x_0} \left(\frac{f_2}{x_2 - x_1} + \frac{f_1}{x_1 - x_2} - \frac{f_1}{x_1 - x_0} - \frac{f_0}{x_0 - x_1} \right)$$

$$= \frac{1}{x_2 - x_0} \left(\frac{f_2 - f_1}{x_2 - x_1} - \frac{f_1 - f_0}{x_1 - x_0} \right).$$

The following example shows how to form divided differences successively by making use of a so-called (divided) difference table.

---

difference table [差分表]

5.4 The Newton forward and backward interpolation formulas

Example 5.2 Use the Newton divided difference interpolation formula to construct a quadratic function p_2 which satisfies

$$p_2(0) = 2 \qquad p_2(1) = -1 \qquad p_2(2) = 4.$$

Solution As in Example 5.1, we denote

$$x_0 = 0, \qquad x_1 = 1, \qquad x_2 = 2$$
$$f_0 = 2, \qquad f_1 = -1, \qquad f_2 = 4.$$

By the Newton divided difference interpolation formula

$$p_2(x) = f_0 + (x - x_0)f[x_0, x_1] + (x - x_0)(x - x_1)f[x_0, x_1, x_2].$$

Now we compute $f[x_0, x_1]$ and $f[x_0, x_1, x_2]$ by the difference table.

x_i	f_i	$f[x_i, x_{i+1}]$	$f[x_i, x_{i+1}, x_{i+2}]$
0	$\underline{2}$		
		$\underline{-3}$	
1	-1		$\underline{4}$
		5	
2	4		

From this table we obtain

$$p_2(x) = 2 - 3(x - x_0) + 4(x - x_0)(x - x_1) = 2 - 3x + 4x(x - 1) = 4x^2 - 7x + 2.$$

$\diamondsuit$

5.4 The Newton forward and backward interpolation formulas

In many applications, the points x_i are given consecutively with equal step-size as

$$x_0, \qquad x_1 = x_0 + h, \qquad x_2 = x_0 + 2h, \qquad \ldots, \qquad x_n = x_0 + nh,$$

where h is a positive number. The function values are

$$f_0 = f(x_0), \qquad f_1 = f(x_0 + h), \qquad f_2 = f(x_0 + 2h), \qquad \ldots, \qquad f_n = f(x_0 + nh).$$

In this case, we can simplify the Newton divided difference interpolation formula.

Define the first forward difference of f at x by

$$\triangle f(x) = f(x+h) - f(x),$$

and the second forward difference of f at x by

$$\triangle^2 f(x) = \triangle(\triangle f(x)) = \triangle f(x+h) - \triangle f(x).$$

Continuing in this way, the mth **forward difference** of f at x is defined by

$$\triangle^m f(x) = \triangle(\triangle^{m-1} f(x)) = \triangle^{m-1} f(x+h) - \triangle^{m-1} f(x).$$

Using the forward difference, we can show

$$f[x_0, \ldots, x_m] = \frac{1}{m! h^m} \triangle^m f(x_0), \qquad (5.7)$$

by induction on m. In fact, it is easily seen that (5.7) holds for $m=1$ because

$$f[x_0, x_1] = \frac{1}{h}(f_1 - f_0) = \frac{1}{1!h} \triangle f(x_0).$$

Assuming (5.7) is true for $m=n$, we show that it holds for $m=n+1$. Since $x_{n+1} = x_0 + (n+1)h$, we get

$$\begin{aligned}
f[x_0, \ldots, x_{n+1}] &= \frac{1}{(n+1)h}(f[x_1, \ldots, x_{n+1}] - f[x_0, \ldots, x_n]) \\
&= \frac{1}{(n+1)h}\left(\frac{1}{n!h^n}\triangle^n f(x_0+h) - \frac{1}{n!h^n}\triangle^n f(x_0)\right) \\
&= \frac{1}{(n+1)!h^{n+1}}\triangle^{n+1} f(x_0),
\end{aligned}$$

which proves (5.7).

Define the binomial coefficients $\begin{pmatrix} \alpha \\ j \end{pmatrix}$ by

$$\begin{pmatrix} \alpha \\ j \end{pmatrix} = \frac{\alpha(\alpha-1)\ldots(\alpha-j+1)}{j!}, \qquad \text{for } j \geq 1$$

---

forward difference [前進差分]   binomial coefficients [二項係数]

5.4 The Newton forward and backward interpolation formulas

and
$$\begin{pmatrix} \alpha \\ 0 \end{pmatrix} = 1.$$

With formula (5.7), the Newton divided difference interpolation formula becomes **Newton's forward difference interpolation formula**

$$\begin{aligned}
p_n(x_0 + \alpha h) &= f_0 + \alpha \triangle f_0 + \frac{\alpha(\alpha-1)}{2!}\triangle^2 f_0 + \cdots + \frac{\alpha(\alpha-1)\cdots(\alpha-n+1)}{n!}\triangle^n f_0 \\
&= \sum_{j=0}^{n} \begin{pmatrix} \alpha \\ j \end{pmatrix} \triangle^j f_0,
\end{aligned} \qquad (5.8)$$

where $\triangle^j f_0 = \triangle^j f(x_0)$.

Instead of forward differences, we may also employ backward differences. We define the first backward difference of f at x by

$$\nabla f(x) = f(x) - f(x-h)$$

the second backward difference of f at x by

$$\nabla^2 f(x) = \nabla f(x) - \nabla f(x-h)$$

and, continuing in this way, the mth **backward difference** of f at x by

$$\nabla^m f(x) = \nabla^{m-1} f(x) - \nabla^{m-1} f(x-h).$$

A formula similar to Newton's forward difference interpolation formula but involving backward difference is **Newton's backward difference interpolation formula**

$$\begin{aligned}
p_n&(x_n + \alpha h) \\
={}&f_n + \alpha \nabla f_n + \frac{\alpha(\alpha+1)}{2!} \nabla^2 f_n + \cdots + \frac{\alpha(\alpha+1)\cdots(\alpha+n-1)}{n!} \nabla^n f_n \\
={}&f_n + (-1)(-\alpha)\nabla f_n + (-1)^2 \frac{-\alpha(-\alpha-1)}{2!} \nabla^2 f_n + \cdots \\
&+ (-1)^n \frac{-\alpha(-\alpha-1)\cdots(-\alpha-n+1)}{n!} \nabla^n f_n \\
={}&\sum_{j=0}^{n}(-1)^j \begin{pmatrix} -\alpha \\ j \end{pmatrix} \nabla^j f_n,
\end{aligned} \qquad (5.9)$$

Newton's forward difference interpolation formula [ニュートンの前進差分補間公式]　backward difference [後退差分]　Newton's backward difference interpolation formula [ニュートンの後退差分補間公式]

where $\nabla^j f_n = \nabla^j f(x_n)$.

Example 5.3 Using (a) Newton's forward formula and (b) Newton's backward formula to construct a polynomial of degree 3 which passes 4 points
$$(0, -5), \quad (1, 1), \quad (2, 9), \quad (3, 25).$$

Solution In this example the step-size is $h = 1$. The computation of the differences is the same in both cases. Only their notation differs.

x_i	f_i	1st Diff	2nd Diff	3rd Diff
0	-5			
		6		
1	1		2	
		8		6
2	9		8	
		16		
3	25			

(a) **Newton's forward formula**   Let $x = x_0 + \alpha h$. Thus $\alpha = (x - x_0)/h = x$. The Newton's forward formula (5.8) gives
$$\begin{aligned} p_3(x) &= -5 + 6x + \frac{2}{2!}x(x-1) + \frac{6}{3!}x(x-1)(x-2) \\ &= x^3 - 2x^2 + 7x - 5. \end{aligned}$$

(b) **Newton's backward formula**   Let $x = x_n + \alpha h$. Thus $\alpha = (x - x_n)/h = x - 3$. The Newton's backward formula (5.9) gives
$$\begin{aligned} p_3(x) &= 25 + 16(x-3) + \frac{8}{2!}(x-3)(x-2) + \frac{6}{3!}(x-3)(x-2)(x-1) \\ &= x^3 - 2x^2 + 7x - 5. \end{aligned} \quad \diamond$$

5.5 Extension of divided differences

Up to now, the divided difference $f[x_0, x_1, \ldots, x_n]$ has been defined for distinct nodes $x_0, x_1, \ldots, x_n$. In this section, we shall extend it for not necessarily distinct nodes. To do so, we state the following theorem.

Theorem 5.1 Let $n \geq 1$ and $x_0, x_1, \ldots, x_n$ be distinct nodes in the interval $[a, b]$. If $f \in C^n[a, b]$, then
$$f[x_0, x_1, \ldots, x_n] = \int_0^{t_0=1} \int_0^{t_1} \cdots \int_0^{t_{n-1}} f^{(n)}(\tau) dt_n \ldots dt_2 dt_1, \quad (5.10)$$

5.5 Extension of divided differences

where
$$\tau = x_0 + \sum_{k=1}^{n} t_k(x_k - x_{k-1}).$$

We note that (5.10) implies
$$t_0 = 1 \geq t_1 \geq \ldots \geq t_{n-1} \geq t_n \geq 0,$$

$$\tau = \sum_{i=0}^{n} \lambda_i x_i,$$

where
$$\lambda_i = t_i - t_{i+1} \geq 0, \qquad 0 \leq i \leq n-1, \qquad \lambda_n = t_n \geq 0$$

and
$$\sum_{i=0}^{n} \lambda_i = 1.$$

Hence
$$a \leq \min\{x_0, x_1, \cdots, x_n\} \leq \tau \leq \max\{x_0, x_1, \cdots, x_n\} \leq b.$$

The right-hand side of (5.10) is well-defined for the case $x_i = x_j$ for some $i \neq j$. Hence, if we redefine $f[x_0, x_1, \ldots, x_n]$ by (5.10), then $f[x_0, x_1, \ldots x_n]$ is a continuous function of $x_0, x_1, \ldots, x_n$. It also keeps the symmetric property
$$f[x_0, x_1, \ldots, x_n] = f[x_{\sigma_0}, x_{\sigma_1}, \ldots, x_{\sigma_n}]$$
for any permutation $\{\sigma_0, \sigma_1, \ldots \sigma_n\}$ of $\{0, 1, 2, \ldots, n\}$. In fact, by the continuity of the divided difference, we have

$$\begin{aligned} f[x_0, x_1, \ldots, x_n] &= \lim_{\substack{\xi_i \to x_i \\ 0 \leq i \leq n}} f[\xi_0, \xi_1, \ldots, \xi_n] \qquad (\xi_i \neq \xi_j,\ i \neq j) \\ &= \lim_{\substack{\xi_i \to x_i \\ 0 \leq i \leq n}} f[\xi_{\sigma_0}, \xi_{\sigma_1}, \ldots, \xi_{\sigma_n}] \\ &= f[x_{\sigma_0}, x_{\sigma_1}, \ldots, x_{\sigma_n}]. \end{aligned}$$

Furthermore, the relation
$$f[x_0, x_1, \ldots, x_n] = \frac{f[x_1, \ldots, x_n] - f[x_0, \ldots, x_{n-1}]}{x_n - x_0}$$

holds if $x_n \neq x_0$, where some of other nodes may be the same.

Proof of Theorem 5.1: The proof is done by induction on n. If $n = 1$, then

$$\int_0^{t_0=1} f'(x_0 + t_1(x_1 - x_0))dt_1 = \left[\frac{f(x_0 + t_1(x_1 - x_0))}{x_1 - x_0}\right]_{t_1=0}^{t_1=1}$$

$$= \frac{f(x_1) - f(x_0)}{x_1 - x_0}$$

$$= f[x_0, x_1].$$

Hence (5.10) holds for $n = 1$. We now assume that (5.10) holds for n distinct nodes. We first observe that

$$\int_0^{t_{n-1}} f^{(n)}\left(x_0 + \sum_{k=1}^n t_k(x_k - x_{k-1})\right) dt_n$$

$$= \left[\frac{f^{(n-1)}(x_0 + \sum_{k=1}^n t_k(x_k - x_{k-1}))}{x_n - x_{n-1}}\right]_{t_n=0}^{t_n=t_{n-1}}$$

$$= \frac{f^{(n-1)}(v) - f^{(n-1)}(u)}{x_n - x_{n-1}},$$

where

$$u = x_0 + \sum_{k=1}^{n-1} t_k(x_k - x_{k-1})$$

$$v = u + t_{n-1}(x_n - x_{n-1}) = x_0 + \sum_{k=1}^{n-2} t_k(x_k - x_{k-1}) + t_{n-1}(x_n - x_{n-2}).$$

Then, the induction assumption implies

$$\int_0^1 \int_0^{t_1} \cdots \int_0^{t_{n-2}} f^{(n-1)}(v) dt_{n-1} \ldots dt_2 dt_1 = f[x_0, x_1, \ldots, x_{n-2}, x_n]$$

and

$$\int_0^1 \int_0^{t_1} \cdots \int_0^{t_{n-2}} f^{(n-1)}(u) dt_{n-1} \ldots dt_2 dt_1 = f[x_0, x_1, \ldots, x_{n-2}, x_{n-1}],$$

5.5 Extension of divided differences

so that

$$\int_0^1 \int_0^{t_1} \cdots \int_0^{t_{n-1}} f^{(n)}(\tau) dt_n \ldots dt_2 dt_1$$
$$= \frac{1}{x_n - x_{n-1}} (f[x_0, x_1, \ldots, x_{n-2}, x_n] - f[x_0, x_1, \ldots, x_{n-2}, x_{n-1}])$$
$$= f[x_0, x_1, \ldots, x_{n-2}, x_{n-1}, x_n].$$

Hence (5.10) holds for $n+1$ distinct nodes. This completes the proof of Theorem 5.4. ∎

Theorem 5.2 *If $f \in C^n[a,b]$, then*

$$f[x_0, x_1, \ldots, x_n] = \frac{1}{n!} f^{(n)}(\xi),$$

where $x_0, x_1, \ldots, x_n$ are not necessarily distinct and

$$\min\{x_0, x_1, \ldots, x_n\} \leq \xi \leq \max\{x_0, x_1, \ldots, x_n\}.$$

Proof: It is easy to see that

$$\int_0^1 \int_0^{t_1} \cdots \int_0^{t_{n-1}} dt_n \ldots dt_2 dt_1 = \frac{1}{n!}.$$

Let

$$\alpha = \min\{x_0, x_1, \ldots, x_n\} \quad \text{and} \quad \beta = \max\{x_0, x_1, \ldots, x_n\},$$

and put

$$m = \min_{\alpha \leq x \leq \beta} f^{(n)}(x) \quad \text{and} \quad M = \max_{\alpha \leq x \leq \beta} f^{(n)}(x).$$

We then have from (5.10)

$$\frac{m}{n!} \leq f[x_0, x_1, \ldots, x_n] \leq \frac{M}{n!}.$$

This implies

$$f[x_0, x_1, \ldots, x_n] = \frac{1}{n!} f^{(n)}(\xi)$$

for some $\xi \in [\alpha, \beta]$. ∎

Corollary 5.1 *If $f \in C^{n+k}[a,b]$, $k \geq 0$, then for any $x \in [a,b]$, the extended divided difference has the property*

$$f[x_0, x_1, \ldots, x_n, \overbrace{x, x, \cdots, x}^{k}] = \frac{1}{(n+k)!} f^{(n+k)}(\eta),$$

where
$$\min\{x_0, x_1, \ldots, x_n, x\} \leq \eta \leq \max\{x_0, x_1, \ldots, x_n, x\}.$$

Proof. This is obtained by applying Theorem 5.2 to the $n+k+1$ nodes $x_0, x_1, \ldots, x_n, x_{n+1} = \cdots = x_{n+k} = x$. ∎

Theorem 5.3 *If $f \in C^{n+k+1}[a,b]$, then for $x \in [a,b]$, we have*

$$\frac{d}{dx} f[x_0, x_1, \ldots, x_n, \overbrace{x, x, \cdots, x}^{k}] = k f[x_0, x_1, \ldots, x_n, \overbrace{x, x, \cdots, x}^{k+1}]$$

and

$$f[x_0, x_1, \ldots, x_n, \overbrace{x, x, \cdots, x}^{k+1}] = \frac{1}{k!} \frac{d^k}{dx^k} f[x_0, x_1, \ldots, x_n, x].$$

Proof: If $k = 1$, then we have

$$\frac{d}{dx} f[x_0, x_1, \ldots, x_n, x]$$
$$= \lim_{\epsilon \to 0} \frac{f[x_0, x_1, \ldots, x_n, x+\epsilon] - f[x_0, x_1, \ldots, x_n, x]}{\epsilon}$$
$$= \lim_{\epsilon \to 0} f[x_0, x_1, \ldots, x_n, x, x+\epsilon]$$
$$= f[x_0, x_1, \ldots, x_n, x, x].$$

If $k = 2$, we have

$$\frac{d}{dx} f[x_0, x_1, \ldots, x_n, x, x]$$
$$= \lim_{\epsilon \to 0} \frac{f[x_0, x_1, \ldots, x_n, x+\epsilon, x+\epsilon] - f[x_0, x_1, \ldots, x_n, x, x]}{\epsilon}$$
$$= \lim_{\epsilon \to 0} \left(\frac{f[x_0, x_1, \ldots, x_n, x+\epsilon, x+\epsilon] - f[x_0, x_1, \ldots, x_n, x, x+\epsilon]}{\epsilon} \right.$$

$$+\frac{f[x_0, x_1, \ldots, x_n, x, x+\epsilon] - f[x_0, x_1, \ldots, x_n, x, x]}{\epsilon}\Bigg)$$
$$= \lim_{\epsilon \to 0} (f[x_0, x_1, \ldots, x_n, x, x+\epsilon, x+\epsilon] + f[x_0, x_1, \ldots x_n, x, x, x+\epsilon])$$
$$= 2f[x_0, x_1, \ldots, x_n, x, x, x].$$

Repeating the argument, we have for all k

$$\frac{d}{dx} f[x_0, x_1, \ldots, x_n, \overbrace{x, x, \cdots, x}^{k}] = k f[x_0, x_1, \ldots, x_n, \overbrace{x, x, \cdots, x}^{k+1}],$$

so that

$$\frac{d^k}{dx^k} f[x_0, x_1, \ldots, x_n, x] = k! f[x_0, x_1, \ldots, x_n, \overbrace{x, x, \cdots, x}^{k+1}].$$

∎

5.6 Error formula

For a given function $f(x)$, we can form an approximation to it by using the interpolation polynomial

$$p_n = a_0 + a_1 x + \ldots + a_n x^n.$$

The two functions f and p_n coincide at $n+1$ points, that is,

$$f(x_i) = p_n(x_i), \quad i = 0, 1, \ldots, n.$$

Using the Newton divided difference interpolation formula in Section 5.3, we obtain

$$\begin{aligned} f(x) &= f(x_0) + (x - x_0) f[x_0, x] \\ f[x_0, x] &= f[x_0, x_1] + (x - x_1) f[x_0, x_1, x] \end{aligned} \quad (5.11)$$

$$\cdots \quad \cdots$$

$$\begin{aligned} f[x_0, \ldots, x_{n-1}, x] &= f[x_0, \ldots, x_n] \\ &\quad + (x - x_n) f[x_0, x_1, \ldots, x_n, x]. \end{aligned} \quad (5.12)$$

By successively substituting from the subsequent relations, we obtain

$$\begin{aligned} f(x) &= f(x_0) + (x - x_0) f[x_0, x_1] \\ &\quad + \ldots + (x - x_0)(x - x_1) \ldots (x - x_{n-1}) f[x_0, x_1, \ldots, x_n] \\ &\quad + (x - x_0)(x - x_1) \ldots (x - x_n) f[x_0, x_1, \ldots, x_n, x] \\ &= p_n(x) + (x - x_0)(x - x_1) \ldots (x - x_n) f[x_0, x_1, \ldots, x_n, x] \end{aligned} \quad (5.13)$$

From (5.13) and Corollary 5.1, we have the following theorem which gives an error formula for the interpolation polynomial p_n.

Theorem 5.4 *Suppose that $x_0, x_1, \ldots, x_n$ are numbers in the interval $[a,b]$ and $f \in C^{n+1}[a,b]$, then for each $x \in [a,b]$, we have*

$$f(x) - p_n(x) = \frac{f^{(n+1)}(\eta)}{(n+1)!} \prod_{i=0}^{n}(x - x_i), \qquad (5.14)$$

where

$$\min\{x_0, x_1, \ldots, x_n, x\} \leq \eta \leq \max\{x_0, x_1, \ldots, x_n, x\}.$$

Example 5.4 Approximate $f(x) = e^x$ using linear interpolation $p_1(x)$ with $x_0 = 0$ and $x_1 = 1$. Find an error bound for $p_1(x)$ on the interval $[x_0, x_1]$.
Solution Applying the Lagrange interpolation formula, we get

$$\begin{aligned} p_1(x) &= \frac{x - x_1}{x_0 - x_1} f(x_0) + \frac{x - x_0}{x_1 - x_0} f(x_1) \\ &= 1 - x + xe \\ &= 1 + (e - 1)x. \end{aligned}$$

By Theorem 5.4, the error in $p_1(x)$ is

$$f(x) - p_1(x) = \frac{f''(\xi(x))}{2}(x - x_0)(x - x_1) = \frac{e^{\xi(x)}}{2} x(x - 1),$$

where $\xi(x) \in [0, 1]$.
For a uniform bound on $[0, 1]$,

$$\max_{0 \leq x \leq 1} e^{\xi(x)} x(1 - x) \leq \max_{0 \leq \xi(x) \leq 1} e^{\xi(x)} \max_{0 \leq x \leq 1} x(1 - x) = \frac{e}{4}.$$

Hence $e/8$ is an error bound for $p_1(x)$ on $[0, 1]$, that is,

$$|f(x) - p_1(x)| \leq \frac{e}{8}, \qquad \text{for } x \in [0, 1]. \qquad \diamond$$

Exercises

(5.1) Use the Lagrange interpolation and the Newton interpolation formulas to construct a quadratic function p_2 which passes through the three points (1,2), (2,6), (3,8).

(5.2) Use Newton's forward and backward formulas to construct a polynomial of degree 3 which passes through the four points (0,0), (1,2), (2, 6), (3,8).

(5.3) Use the Lagrange interpolation or the Newton interpolation formulas to construct interpolation polynomials for the following functions and find a bound for the absolute error on the interval $[x_0, x_n]$.

(5.3.1) $f(x) = e^x$, $x_0 = 0,\ x_1 = 0.5,\ x_2 = 1.0$

(5.3.2) $f(x) = \sin x$, $x_0 = 0,\ x_1 = 0.2,\ x_2 = 0.4$

(5.3.1) $f(x) = \ln x$, $x_0 = 1,\ x_1 = 1.2,\ x_2 = 1.3$

(5.4) Show that the Lagrange factors satisfy

$$\sum_{k=0}^{n} \ell_k(x) = 1.$$

(5.5) Let $f(x)$ be a polynomial of degree $\leq n$. Let $p_n(x)$ be the Lagrange interpolation polynomial

$$p_n(x) = \sum_{k=0}^{n} f(x_k)\ell_k(x),$$

where $x_0 < x_1 < \cdots < x_n$. Show that $f(x) = p_n(x)$ for all x.

Chapter 6.

Numerical Integration

6.1 Introduction

The problem of numerical integration is to compute approximate values of the definite integral

$$I(f) = \int_a^b f(x)dx,$$

by a sum of the type

$$I_n(f) = \alpha_0 f(x_0) + \ldots + \alpha_n f(x_n), \qquad (6.1)$$

where

$$a = x_0 < x_1 < \ldots < x_n = b.$$

The sum (6.1) is called numerical quadrature. The coefficients α_i are called the integration weights or quadrature weights, and the points are the integration nodes, usually chosen in $[a, b]$.

If we can find a differentiable function F whose derivative is f, then we can evaluate $I(f)$ by applying the formula

$$I(f) = \int_a^b f(x)dx = F(b) - F(a).$$

---

numerical integration [数値積分]　　definite integral [定積分]　　quadrature [積分、求積]　　integration weights [数値積分の重み]

6.1 Introduction

In practice, however, f often has no explicit antiderivative or its antiderivative is not easily obtained. Then we need a numerical method of approximate integration.

The methods of numerical integration in this chapter are based on the interpolation polynomials described in Chapter 5. Let p_n be the Lagrange interpolation polynomial of f

$$p_n(x) = \sum_{i=0}^{n} f(x_i) \ell_i(x).$$

By Theorem 5.4, if $f \in C^{n+1}[a, b]$, f can be expressed as

$$\begin{aligned} f(x) &= p_n(x) + f[x_0, x_1, \ldots, x_n, x] \prod_{i=0}^{n} (x - x_i) \\ &= p_n(x) + \frac{f^{(n+1)}(\eta(x))}{(n+1)!} \prod_{i=0}^{n} (x - x_i), \end{aligned}$$

where $\eta(x)$ is in $[a, b]$.

We integrate f, p_n and the truncation error term over $[a, b]$ to obtain

$$\begin{aligned} & \int_a^b f(x) dx \\ &= \int_a^b p_n(x) dx + \int_a^b f[x_0, x_1, \ldots, x_n, x] \prod_{i=0}^{n} (x - x_i) dx \\ &= \sum_{i=0}^{n} f(x_i) \int_a^b \ell_i(x) dx + \int_a^b f[x_0, x_1, \ldots, x_n, x] \prod_{i=0}^{n} (x - x_i) dx \\ &= \sum_{i=0}^{n} \alpha_i f(x_i) + \frac{1}{(n+1)!} \int_a^b f^{(n+1)}(\eta(x)) \prod_{i=0}^{n} (x - x_i) dx \end{aligned}$$

where

$$\alpha_i = \int_a^b \ell_i(x) dx, \qquad i = 0, 1, \ldots, n. \tag{6.2}$$

Therefore, the quadrature formula is

$$\int_a^b f(x) dx \approx I_n(f) = \sum_{i=0}^{n} \alpha_i f(x_i), \tag{6.3}$$

---

antiderivative [原始関数]

with error given by

$$E_n(f) = I(f) - I_n(f) = \int_a^b f[x_0, x_1, \ldots, x_n, x] \prod_{i=0}^n (x - x_i) dx$$

$$= \frac{1}{(n+1)!} \int_a^b \prod_{i=0}^n (x - x_i) f^{(n+1)}(\eta(x)) dx.$$

The formula (6.3) with (6.2) is called the **Newton-Cotes quadrature formula** of order n.

If the degree n of the polynomial is too high, some of the coefficients α_i become negative and $|\alpha_i| \to \infty$ as $n \to \infty$. In this case, errors due to round-off and local instability can cause a problem. This is why we often use Newton-Cotes methods which divide the interval $[a,b]$ into n subintervals of equal length, and integrate a lower-degree polynomial approximation of f on each subinterval. In this chapter, we will study Newton-Cotes methods using polynomials of degrees 0, 1 and 2, which are the rectangular rule, the trapezoidal rule and the Simpson rule, respectively. To estimate the error of the three rules, we need the **integral mean value theorem**.

Theorem 6.1 *Let $r(x)$ be constant sign (i.e., nonnegative or nonpositive) and integrable on $[a,b]$, and let $q(x)$ be continuous on $[a,b]$. Then*

$$\int_a^b q(x) r(x) dx = q(\xi) \int_a^b r(x) dx,$$

for some $\xi \in [a,b]$.

Proof: Let

$$\underline{q} = \min_{a \le x \le b} q(x) \quad \text{and} \quad \bar{q} = \max_{a \le x \le b} q(x).$$

If $r(x) \ge 0$ in $[a,b]$, then

$$\underline{q} r(x) \le q(x) r(x) \le \bar{q} r(x), \qquad x \in [a,b].$$

Integrating from a to b, we have

$$\underline{q} \int_a^b r(x) dx \le \int_a^b q(x) r(x) dx \le \bar{q} \int_a^b r(x) dx.$$

Newton-Cotes quadrature formula [ニュートン・コーツ積分公式] rectangular rule [矩形 (長方形) 則] trapezoidal rule [台形則] integral mean value theorem [積分の平均値定理]

6.2 The rectangular rule

This implies
$$\int_a^b q(x)r(x)dx = q(\xi) \int_a^b r(x)dx,$$
for some $\xi \in [a,b]$. A similar argument works for the case $r(x) \leq 0$ in $[a,b]$. ∎

6.2 The rectangular rule

The simplest formula is the **rectangular rule** which is produced by using the polynomial interpolation of degree 0.

We divide the interval $[a,b]$ into n subintervals $[a_i, b_i]$ of equal length $h = \dfrac{b-a}{n}$, where
$$a_i = a + ih, \qquad b_i = a + (i+1)h = a_i + h, \qquad i = 0, \ldots, n-1.$$
In each subinterval $[a_i, b_i]$, we approximate f by
$$p_{0,i}(x) = f\left(\frac{a_i + b_i}{2}\right),$$
that is, the value of f at the midpoint $\dfrac{a_i + b_i}{2}$ of $[a_i, b_i]$. Then the formula
$$\int_{a_i}^{b_i} f(x)dx \approx hf\left(\frac{a_i + b_i}{2}\right)$$
is called the **midpoint formula**.

Application of the midpoint formula to n subintervals $[a_i, b_i], i = 0, \ldots, n-1$ yields
$$I(f) = \int_a^b f(x)dx = \sum_{i=0}^{n-1} \int_{a_i}^{b_i} f(x)dx \approx h \sum_{i=0}^{n-1} f\left(\frac{a_i + b_i}{2}\right).$$
The form
$$R_n(f) = h \sum_{i=0}^{n-1} f\left(\frac{a_i + b_i}{2}\right)$$
is called the **rectangular rule** or **the compound midpoint rule** (See Figure 6.1).

---

midpoint formula [中点公式, 中点則]　　compound midpoint rule [複合中点則, 合成中点則]

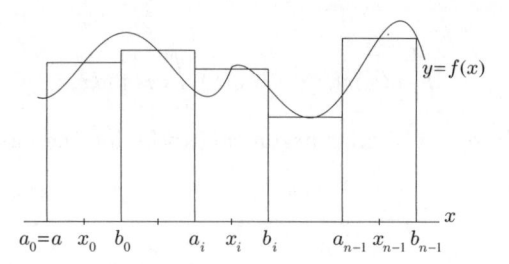

Figure 6.1: The rectangular rule

In order to estimate the error of this rule, let

$$x_i = \frac{a_i + b_i}{2} \quad \text{and} \quad r(x) = \int_{a_i}^{x} (t - x_i)dt.$$

Then $r(x) \leq 0$ in $[a_i, b_i]$ and we obtain

$$\begin{aligned}
\int_{a_i}^{b_i} f(x)dx - hf(x_i) &= \int_{a_i}^{b_i} (f(x) - f(x_i))dx \\
&= \int_{a_i}^{b_i} (x - x_i) f[x_i, x] dx \\
&= \int_{a_i}^{b_i} r'(x) f[x_i, x] dx \\
&= [r(x) f[x_i, x]]_{a_i}^{b_i} - \int_{a_i}^{b_i} r(x) \frac{d}{dx} f[x_i, x] dx \\
&= -\int_{a_i}^{b_i} r(x) f[x_i, x, x] dx \\
&= -f[x_i, \xi_i, \xi_i] \int_{a_i}^{b_i} r(x) dx \\
&= -\frac{f''(\lambda_i)}{2} \int_{a_i}^{b_i} r(x) dx \\
&= -\frac{f''(\lambda_i)}{4} \int_{a_i}^{b_i} [(x - x_i)^2 - (a_i - x_i)^2] dx \\
&= \frac{h^3 f''(\lambda_i)}{24}
\end{aligned}$$

where $\xi_i, \lambda_i \in [a_i, b_i]$. We have used Corollary 5.1, Theorem 6.1 and Theorem 5.3 for the 5th, 6th, 7th equalities.

6.2 The rectangular rule

The reason for introducing $r(x)$ is that $(x - x_i)$ in $[a_i, b_i]$ changes sign at x_i, and we cannot apply the integral mean value theorem to

$$\int_{a_i}^{b_i} (x - x_i) f[x_i, x] dx.$$

Hence the error of the rectangular rule is given by

$$\int_a^b f(x)dx - R_n(f) = \frac{h^3}{24} \sum_{i=0}^{n-1} f''(\lambda_i) = \frac{h^2(b-a)}{24} f''(\lambda), \qquad (6.4)$$

where $\lambda_i \in [a_i, b_i]$, and $\lambda \in [a, b]$ satisfies

$$\frac{1}{n} \sum_{i=0}^{n-1} f''(\lambda_i) = f''(\lambda).$$

Example 6.1 Evaluate

$$I(f) = \int_0^1 e^{-x^2} dx$$

by the rectangular rule with $n = 10$.

Solution Dividing the interval $[0, 1]$ into 10 subintervals, we get

$$h = \frac{1}{10}, \qquad [a_i, b_i] = [ih, (i+1)h], \qquad i = 0, 1, \ldots, 9.$$

The midpoints of $[a_i, b_i]$ are

$$x_i = \frac{a_i + b_i}{2} = \frac{1}{2}h + ih = \frac{1}{20} + ih = 0.05 + ih, \qquad i = 0, 1, \ldots, 9.$$

Then, evaluating $I(f)$ by the rectangular rule with $n = 10$ gives

$$\begin{aligned} R_{10}(f) &= h \sum_{i=0}^{9} f(x_i) \\ &= 0.1[f(0.05) + f(0.15) + f(0.25) + f(0.35) + f(0.45) \\ &\quad + f(0.55) + f(0.65) + f(0.75) + f(0.85) + f(0.95)] \\ &= 0.747131. \end{aligned}$$

---

sign [符号]

For the error,

$$\begin{aligned}|E_{10}(f)| &= |I(f) - R_{10}(f)| \\ &\leq \frac{h^2}{24}|f''(\lambda)| \\ &= \frac{h^2}{24}|(2\lambda)^2 - 2|e^{-\lambda^2} \\ &\leq \frac{h^2}{12} = 0.00083,\end{aligned}$$

where we have used $\lambda \in [0,1]$ and $|(2\lambda)^2 - 2|e^{-\lambda^2} \leq 2$. $\diamond$

6.3 The trapezoidal rule

The trapezoidal rule is produced by using the Lagrange polynomial of degree 1.

The first step is similar to the rectangular rule. We divide the interval $[a,b]$ into n subintervals $[a_i, b_i]$ of equal length $h = (b-a)/n$

$$a_i = a + ih, \qquad b_i = a_i + h, \qquad i = 0, \ldots, n-1.$$

To derive the trapezoidal rule for approximating $I(f)$, in each subinterval $[a_i, b_i]$, we approximate f by the line segment (linear interpolation polynomial) passing through the two points $(a_i, f(a_i))$ and $(b_i, f(b_i))$, that is,

$$\begin{aligned}p_{1,i}(x) &= \frac{x-b_i}{a_i-b_i}f(a_i) + \frac{x-a_i}{b_i-a_i}f(b_i) \\ &= \frac{1}{h}[(b_i-x)f(a_i) + (x-a_i)f(b_i)].\end{aligned}$$

Thus

$$\begin{aligned}\int_{a_i}^{b_i} f(x)dx &\approx \int_{a_i}^{b_i} p_{1,i}(x)dx \\ &= \frac{1}{2h}\left[-(x-b_i)^2 f(a_i) + (x-a_i)^2 f(b_i)\right]_{a_i}^{b_i} \\ &= \frac{h}{2}[f(a_i) + f(b_i)].\end{aligned}$$

Obviously, $a_i = b_{i-1}$, $i = 1, 2, \ldots, n-1$. We set

$$x_0 = a, \; x_n = b, \; x_i = a_i = b_{i-1}, \qquad i = 1, 2, \ldots, n-1.$$

6.3 The trapezoidal rule

By summing all these formulas, we obtain

$$\begin{aligned} I(f) &= \int_a^b f(x)dx \\ &= \sum_{i=0}^{n-1} \int_{a_i}^{b_i} f(x)dx \\ &\approx \sum_{i=0}^{n-1} \frac{h}{2}[f(a_i) + f(b_i)] \\ &= h\left[\frac{1}{2}f(a) + f(x_1) + f(x_2) + \ldots + f(x_{n-1}) + \frac{1}{2}f(b)\right]. \end{aligned}$$

The trapezoidal rule (or compound trapezoidal rule) is of the form (See Figure 6.2)

$$T_n(f) = h\left[\frac{1}{2}f(a) + f(x_1) + f(x_2) + \ldots + f(x_{n-1}) + \frac{1}{2}f(b)\right].$$

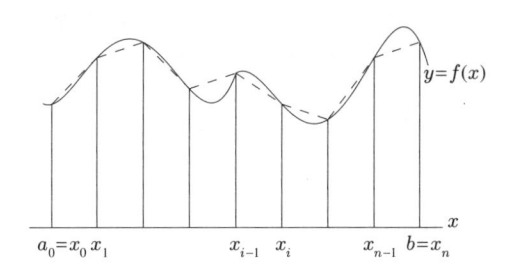

Figure 6.2: The trapezoidal rule

To obtain an error formula, we apply Theorem 5.4, Theorem 5.2 and Theorem 6.1 to each subinterval and find

$$\begin{aligned} \int_{a_i}^{b_i} f(x)dx - \frac{h}{2}[f(a_i) + f(b_i)] &= \int_{a_i}^{b_i} [f(x) - p_{1,i}(x)]dx \\ &= \int_{a_i}^{b_i} (x - a_i)(x - b_i)f[a_i, b_i, x]dx \\ &= f[a_i, b_i, \eta_i] \int_{a_i}^{b_i} (x - a_i)(x - b_i)dx \end{aligned}$$

---

the trapezoidal rule [台形則]   compound trapezoidal rule [合成台形則]

$$\begin{aligned} &= \frac{1}{2}f''(\mu_i)\int_{a_i}^{b_i}(x-a_i)(x-b_i)dx \\ &= \frac{1}{2}f''(\mu_i)\left[-\frac{1}{6}(b_i-a_i)^3\right], \end{aligned}$$

where $\eta_i, \mu_i \in [a_i, b_i]$. Hence for the error in the trapezoidal rule,

$$\begin{aligned} E_n(f) &= \int_a^b f(x)dx - T_n(f) \\ &= -\frac{h^3}{12}\sum_{i=0}^{n-1} f''(\mu_i) \quad (6.5) \\ &= -\frac{h^2(b-a)}{12}f''(\mu), \end{aligned}$$

where $\mu \in [a, b]$.

Remark Comparing the error of the rectangular rule (6.4) with (6.5), the rectangular rule seems to yield rather sharper result than the trapezoidal rule. However, the values of λ_i and μ_i in the two rules are different and they have also significant impact on the error. Moreover, it can be shown that the trapezoidal rule has the error

$$E(f) = \frac{h^2}{24}(f'(b)-f'(a)) - \frac{7h^4}{5760}(f'''(b)-f'''(a)) + \dots.$$

(See [Yamamoto, 1976]). This means that the trapezoidal rule is quite suitable for periodic functions.

Example 6.2 Evaluate

$$I(f) = \int_0^1 e^{-x^2}dx$$

by the trapezoidal rule with $n = 10$.

Solution Dividing the interval $[0, 1]$ into 10 subintervals, we get

$$h = \frac{1}{10}, \quad [a_i, b_i] = [ih, (i+1)h], \quad i = 0, 1, \dots, 9.$$

The endpoints of $[a_i, b_i]$ are

$$a = 0, \quad x_i = ih, \quad i = 1, \dots, 9, \quad b = 1.$$

---

sharper [sharp(鋭い, 良い) の比較級]   periodic function [周期関数]

Then evaluating $I(f)$ by the trapezoidal rule with $n = 10$ gives

$$\begin{aligned}T_{10}(f) &= 0.1\left[\frac{1}{2}f(0) + f(0.1) + f(0.2) + f(0.3) + f(0.4)\right.\\ &\quad \left. + f(0.5) + f(0.6) + f(0.7) + f(0.8) + f(0.9) + \frac{1}{2}f(1)\right]\\ &= 0.746212.\end{aligned}$$

For the error,

$$|E_{10}(f)| \leq \frac{h^2}{12}|f''(\mu)| = \frac{h^2}{12}|(2\mu)^2 - 2|e^{-\mu^2} \leq \frac{h^2}{6} \approx 0.0017. \quad \diamond$$

6.4 Simpson's rule

The rectangular rule is produced by the constant interpolating polynomial p_0 of f and the trapezoidal rule is produced by linear interpolating polynomial p_1. We will show that the Simpson rule is produced by the quadratic interpolating polynomial p_2.

The first step is similar to the rectangular rule and the trapezoidal rule. We divide the interval $[a, b]$ into n subintervals $[a_i, b_i]$ of equal length $h = (b-a)/n$

$$a_i = a + ih, \qquad b_i = a_i + h, \qquad i = 0, \ldots, n-1.$$

The Simpson rule differs from the rectangular rule and the trapezoidal rule in that the Simpson rule n **must be an even number**.

To derive the Simpson rule for approximating $I(f)$, we set the endpoints of these subintervals by

$$a = x_0, \quad b_0 = a_1 = x_1, \quad b_1 = a_2 = x_2, \quad \ldots, \quad b = x_n.$$

We approximate f in subintervals

$$[x_i, x_{i+1}] \cup [x_{i+1}, x_{i+2}], \qquad i = 0, 2, 4, \ldots, n-2,$$

---

endpoints[端点]

by the Lagrange interpolation polynomial of order 2,

$$\begin{aligned}
p_{2,i}(x) &= \frac{(x - x_{i+1})(x - x_{i+2})}{(x_i - x_{i+1})(x_i - x_{i+2})} f(x_i) + \frac{(x - x_i)(x - x_{i+2})}{(x_{i+1} - x_i)(x_{i+1} - x_{i+2})} f(x_{i+1}) \\
&\quad + \frac{(x - x_i)(x - x_{i+1})}{(x_{i+2} - x_i)(x_{i+2} - x_{i+1})} f(x_{i+2}) \\
&= \frac{(x - x_{i+1})(x - x_{i+2})}{2h^2} f(x_i) + \frac{(x - x_i)(x - x_{i+2})}{-h^2} f(x_{i+1}) \\
&\quad + \frac{(x - x_i)(x - x_{i+1})}{2h^2} f(x_{i+2}).
\end{aligned}$$

Thus

$$\int_{x_i}^{x_{i+2}} f(x)dx = \int_{x_i}^{x_{i+1}} f(x)dx + \int_{x_{i+1}}^{x_{i+2}} f(x)dx \approx \int_{x_i}^{x_{i+2}} p_{2,i}(x)dx.$$

Setting $s = (x - x_{i+1})/h$, we have

$$dx = hds, \qquad x - x_i = (s+1)h, \qquad x - x_{i+1} = sh, \qquad x - x_{i+2} = (s-1)h,$$

and obtain

$$\begin{aligned}
\int_{x_i}^{x_{i+2}} p_{2,i}(x)dx &= h \int_{-1}^{1} \left[\frac{1}{2} s(s-1) f(x_i) - (s+1)(s-1) f(x_{i+1}) + \frac{1}{2}(s+1)s f(x_{i+2}) \right] ds \\
&= h \left[\frac{1}{3} f(x_i) + \frac{4}{3} f(x_{i+1}) + \frac{1}{3} f(x_{i+2}) \right].
\end{aligned}$$

By summing all these formulas $i = 0, 2, 4, \ldots, n-2$, we obtain

$$\begin{aligned}
I(f) &= \int_a^b f(x)dx = \sum_{i=1}^{n/2} \int_{x_{2i-2}}^{x_{2i}} f(x)dx \\
&\approx \frac{h}{3} [f(a) + 4f(x_1) + 2f(x_2) + \ldots + 2f(x_{n-2}) + 4f(x_{n-1}) + f(b)].
\end{aligned}$$

The Simpson rule is of the form

$$S_n(f) = \frac{h}{3} [f(a) + f(b) + 4(f(x_1) + f(x_3) + \ldots + f(x_{n-1}))$$

6.4 Simpson's rule

$$+2(f(x_2) + f(x_4) + \ldots + f(x_{n-2}))].$$

To estimate the error in the Simpson rule, we introduce the function

$$r(x) = \int_{x_i}^{x} (t - x_i)(t - x_{i+1})(t - x_{i+2})dt,$$

since the polynomial $(x - x_i)(x - x_{i+1})(x - x_{i+2})$ changes sign at x_{i+1} (See Figure 6.3). It is easy to verify

$$r(x_i) = r(x_{i+2}) = 0, \quad r(x) > 0 \quad \text{for } x_i < x < x_{i+2}.$$

Then by Theorem 5.4, we obtain

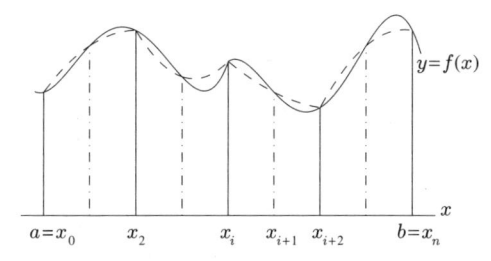

Figure 6.3: The Simpson rule

$$\int_{x_i}^{x_{i+2}} (f(x) - p_{2,i}(x))dx$$

$$= \int_{x_i}^{x_{i+2}} (x - x_i)(x - x_{i+1})(x - x_{i+2})f[x_i, x_{i+1}, x_{i+2}, x]dx$$

$$= \int_{x_i}^{x_{i+2}} r'(x)f[x_i, x_{i+1}, x_{i+2}, x]dx$$

$$= [r(x)f[x_i, x_{i+1}, x_{i+2}, x]]_{x_i}^{x_{i+2}} - \int_{x_i}^{x_{i+2}} r(x)\frac{d}{dx}f[x_i, x_{i+1}, x_{i+2}, x]dx$$

$$= -\int_{x_i}^{x_{i+2}} r(x)f[x_i, x_{i+1}, x_{i+2}, x, x]dx$$

$$= -f[x_i, x_{i+1}, x_{i+2}, \xi_i, \xi_i] \int_{x_i}^{x_{i+2}} r(x)dx$$

$$= -\frac{f^{(4)}(\nu_i)}{4!} \int_{x_i}^{x_{i+2}} r(x)dx$$

$$= -\frac{h^4 f^{(4)}(\nu_i)}{24} \int_{x_i}^{x_{i+2}} \int_{-1}^{(x-x_{i+1})/h} (s+1)s(s-1)dsdx$$

$$= -\frac{h^4 f^{(4)}(\nu_i)}{24}\left[\frac{4}{15}h\right]$$
$$= -\frac{h^5}{90}f^{(4)}(\nu_i),$$

where $\xi_i, \nu_i \in [x_i, x_{i+2}]$.

Hence the error of the Simpson rule is given by

$$\begin{aligned}
E_n(f) &= \int_a^b f(x)dx - S_n(f) \\
&= -\sum_{i=0}^{n/2} \frac{h^5}{90} f^{(4)}(\nu_i) \\
&= -\frac{h^4}{90}\frac{b-a}{n}\sum_{i=0}^{n/2} f^{(4)}(\nu_i) \\
&= -\frac{h^4(b-a)}{180} f^{(4)}(\nu),
\end{aligned} \qquad (6.6)$$

where $\nu_i \in [x_{2i-2}, x_{2i}]$ and $\nu \in [a,b]$.

Example 6.3 Evaluate

$$I(f) = \int_0^1 e^{-x^2} dx$$

by the Simpson rule with $n = 10$.

Solution Dividing the interval $[0,1]$ into 10 subintervals, we get

$$h = \frac{1}{10}, \quad [a_i, b_i] = [ih, (i+1)h], \quad i = 0, 1, \ldots, 9.$$

The endpoints of $[a_i, b_i]$ are

$$a = 0, \quad x_i = ih, \quad i = 1, \ldots, 9, \quad b = 1.$$

Then evaluating $I(f)$ by the Simpson rule with $n = 10$ gives

$$\begin{aligned}
S_{10}(f) &= \frac{0.1}{3}[f(0) + 4f(0.1) + 2f(0.2) + 4f(0.3) + 2f(0.4) \\
&\quad + 4f(0.5) + 2f(0.6) + 4f(0.7) + 2f(0.8) + 4f(0.9) + f(1)] \\
&\approx 0.746825.
\end{aligned}$$

For the error

$$|E_{10}(f)| \leq \frac{h^4}{180}|f^{(4)}(\nu)| = \frac{h^4}{45}|4\nu^4 - 12\nu^2 + 3|e^{-\nu^2} \leq \frac{h^4}{15} \approx 0.000007,$$

where we have used

$$f^{(4)}(x) = 4(4x^4 - 12x^2 + 3)e^{-x^2}$$

and an estimate

$$|4x^4 - 12x^2 + 3|e^{-x^2} \leq 3, \qquad 0 \leq x \leq 1. \qquad \diamondsuit$$

Note: QUAD routine in Matlab for numerical integration uses the Simpson's rule, which gives

$$I(f) = \int_0^1 e^{-x^2} dx \approx 0.746826.$$

Exercises

(6.1) Approximate $\int_0^1 x \sin x \, dx$ by using the following three rules and step sizes:

(i) Rectangular rule with $h = 0.2$ and $h = 0.1$
(ii) Trapezoidal rule with $h = 0.2$ and $h = 0.1$
(iii) Simpson's rule with $h = 0.25$ and $h = 0.1$.

Compare the approximation with the true value $\sin(1) - \cos(1)$.

(6.2) Estimate absolute errors for the approximate values computed by the three rules in Exercise (6.1).

(6.3) Apply Simpson's rule to compute $\int_1^2 \frac{1}{x} dx$ with $h = 0.5$.

Show the following inequality:

$$\left| \int_1^2 \frac{1}{x} dx - I_2\left(\frac{1}{x}\right) \right| \leq \int_1^2 \left|(x-1)\left(x - \frac{3}{2}\right)(x-2)\right| dx.$$

(6.4) Divide the interval $[0, 1]$ into 10 and 20 subintervals of equal length and approximate $\int_0^1 \frac{1}{1+x^2} dx$ by using the following three rules:

(i) Rectangular rule

(ii) Trapezoidal rule

(iii) Simpson's rule

Compare the approximation with the true value $\pi/4$.

(6.5) Determine the value of h required to approximate the integral in Exercises (6.1) and (6.4) to within 10^{-2}.

Chapter 7.

Initial Value Problems for Ordinary Differential Equations

7.1 Introduction

This chapter is concerned with approximating the solution $y(x)$ of the first order ordinary differential equation

$$\frac{dy}{dx} = f(x, y), \qquad a \leq x \leq b, \tag{7.1}$$

subject to the initial condition

$$y(a) = y_0. \tag{7.2}$$

The following theorem is a version of the fundamental existence and uniqueness theorem for the initial value problem (7.1)–(7.2).

Theorem 7.1 *Suppose that $f(x, y)$ is continuous on*

$$D = \{(x, y) \mid a \leq x \leq b, -\infty < y < \infty\}.$$

---

initial value problem [初期値問題]   ordinary differential equation [常微分方程式]
first order [1 階]   existence [存在]   uniqueness [一意性]

102 Chapter 7. Initial Value Problems for Ordinary Differential Equations

If f satisfies a Lipschitz condition on D with respect to the second variable, that is, there is a positive constant K such that

$$|f(x,y) - f(x,z)| \leq K|y - z|, \qquad \text{for } (x,y), (x,z) \in D,$$

then the initial-value problem (7.1) - (7.2) has a unique solution $y(x)$ for $a \leq x \leq b$.

The proof of Theorem 7.1 can be found in most textbooks on the theory of ordinary differential equations, or see [Kress 1998].

In general, it is difficult to find an exact solution $y(x)$ of the initial value problem. Numerical methods for the initial value problem find approximate values of $y(x_i)$, $i = 1, 2, \ldots n$ at the points

$$x_0 = a, \qquad x_i = x_0 + ih, \qquad i = 1, 2, \ldots, n, \qquad h = \frac{b-a}{n}.$$

The points $x_0, x_1, x_2, \ldots, x_n$ are called **mesh points** and h is called the **step size**.

These methods considered in this chapter are step by step methods, that is, we start from the given initial value $y_0 = y(x_0)$ and proceed stepwise. Thus, the error grows as the value of x_i increases. If small changes or errors at $y(x_i)$ introduce correspondingly small changes at $y(x)$ for $x \geq x_i$, we say the initial value problem is a **well posed problem** or **stable problem**. Otherwise, we say the problem is an **ill posed problem** or **unstable problem**. It is known that if f satisfies conditions of Theorem 7.1, then the initial value problem is well posed.

7.2 Euler's method

Suppose that $y(x)$, the unique solution of the initial value problem (7.1) - (7.2), is twice continuously differentiable on $[a,b]$. Using the Taylor expansion, for each $i = 0, 1, \ldots, n-1$, we have

$$y(x_{i+1}) = y(x_i) + (x_{i+1} - x_i)y'(x_i) + \frac{(x_{i+1} - x_i)^2}{2}y''(\xi_i),$$

---

mesh points [分点]　　step size [刻み幅]　　well posed problem [良条件問題]　　stable problem [安定な問題]　　ill posed problem [悪条件問題]　　unstable problem [不安定な問題]

7.2 Euler's method

where $\xi_i \in (x_i, x_{i+1})$. Since $x_{i+1} - x_i = h$ and y satisfies (7.1) - (7.2),

$$y(x_{i+1}) = y(x_i) + hf(x_i, y(x_i)) + \frac{h^2}{2}y''(\xi_i). \qquad (7.3)$$

Now, for a small step size h, its square h^2 is relatively small. This suggests the approximation

$$y(x_{i+1}) \approx y(x_i) + hf(x_i, y(x_i)), \qquad (7.4)$$

which leads to **Euler's method**

$$Y_{i+1} = Y_i + hf(x_i, Y_i), \qquad i = 0, 1, \ldots, n-1.$$

In this method, we first compute

$$Y_1 = y_0 + hf(x_0, Y_0)$$

which approximates $y(x_1)$. Then we compute

$$Y_2 = Y_1 + hf(x_1, Y_1)$$

which approximates $y(x_2)$, etc. See Figure 7.1.

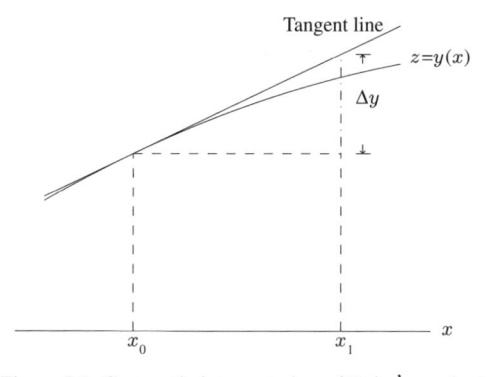

Figure 7.1: Geometric interpretation of Euler's method

The Euler method is called a **first order method**, because the approximation (7.4) takes only the constant term and the term containing the first power of h in the Taylor expansion (7.3). The local truncation error

---

first order method [1 次の方法] (p.101 の first order と異なる訳に注意)   local truncation error [局所打切り誤差]   global truncation error [大域打切り誤差]

τ_i at x_i and the global truncation error τ for Euler's method are then defined by
$$\tau_i = \frac{y(x_{i+1}) - [y(x_i) + hf(x_i, y(x_i))]}{h}$$
and
$$\tau = \max_{1 \leq i \leq n-1} |\tau_i|.$$
Hence we have $\tau = O(h)$.

In addition, there are round-off errors in this method, which may affect the accuracy of the approximate values $Y_1, Y_2, \ldots, Y_n$ more and more as i increases. The error in Y_i can be estimated if f satisfies the Lipschitz condition of Theorem 7.1 and $|y''(x)| \leq M$ for $x \in [a, b]$, where M is a positive constant. Denote the error in Y_i by
$$e_i = y(x_i) - Y_i, \qquad 0 \leq i \leq n.$$
Then from
$$y(x_{i+1}) = y(x_i) + hf(x_i, y(x_i)) + \frac{1}{2}h^2 y''(\xi_i)$$
and
$$Y_{i+1} = Y_i + hf(x_i, Y_i),$$
we get
$$\begin{aligned} |e_{i+1}| &\leq |e_i| + h|f(x_i, y(x_i)) - f(x_i, Y_i)| + \frac{1}{2}h^2 |y''(\xi_i)| \\ &\leq |e_i| + hK|y(x_i) - Y_i| + \frac{1}{2}h^2 M \\ &= (1 + hK)|e_i| + \frac{1}{2}h^2 M \\ &\leq (1 + hK)\left((1 + hK)|e_{i-1}| + \frac{1}{2}h^2 M\right) + \frac{1}{2}h^2 M \\ &\leq (1 + hK)^{i+1}|e_0| + \frac{1}{2}h^2 M \sum_{j=0}^{i}(1 + hK)^j. \end{aligned}$$
Since $|e_0| = 0$ and
$$\sum_{j=0}^{i} hK(1 + hK)^j = (1 + hK)^{i+1} - 1 < e^{hK(i+1)} - 1 \leq e^{K(b-a)} - 1,$$

7.3 Runge-Kutta methods

we obtain the error bound

$$|e_{i+1}| \leq \frac{hM}{2K}(e^{K(b-a)} - 1),$$

that is,

$$\max_i |y(x_i) - Y_i| = O(h).$$

Example 7.1 Apply the Euler method with $h = 0.1$ to the following initial value problem

$$\frac{dy}{dx} = x + y, \qquad 0 \leq x \leq 1$$
$$y(0) = 0.$$

Solution The mesh points are

$x_0 = 0$, $x_1 = x_0 + h = 0.1$, $x_2 = x_0 + 2h = 0.2$, ..., $x_{10} = x_0 + nh = 1$.

Let $f(x, y) = x + y$. Then applying the Euler method to this problem gives

$$Y_{i+1} = Y_i + 0.1(x_i + Y_i), \qquad i = 0, 1, 2, \ldots, 9.$$

We compute Y_i step by step, and obtain

$Y_1 = 0.0$, $Y_2 = 0.0100$, $Y_3 = 0.0310$, $Y_4 = 0.0641$, $Y_5 = 0.1105$

$Y_6 = 0.1716$, $Y_7 = 0.2487$, $Y_8 = 0.3436$, $Y_9 = 0.4579$, $Y_{10} = 0.5937$. $\diamond$

The problem in Example 7.1 satisfies the Lipschitz condition of Theorem 7.1 with $K = 1$, and its solution is $y(x) = e^x - x - 1$, whose second derivative satisfies $|y''(x)| = e^x \leq e$ for $x \in [0, 1]$. Hence the problem is well-posed, and the error is

$$\max_i |y(x_i) - Y_i| \leq \frac{e}{2}(e - 1)h.$$

7.3 Runge-Kutta methods

Runge-Kutta methods have higher order local truncation error than the Euler method. The Runge-Kutta methods are based on Taylor expansions for the functions $y(x)$ and $f(x, y(x))$. We will describe two important Runge-Kutta methods which are the second-order Runge-Kutta method and the fourth-order Runge-Kutta method.

7.3.1 Second-order Runge-Kutta method

Expanding $y(x)$ in terms of its Taylor polynomial about x gives

$$y(x+h) = y(x) + hy'(x) + \frac{h^2}{2}y''(x) + O(h^3).$$

Since $y'(x) = f(x, y(x))$, we obtain

$$y(x+h) = y(x) + hf(x, y(x)) + \frac{h^2}{2}\frac{df}{dx}(x, y(x)) + O(h^3). \qquad (7.5)$$

If we expand $f(x, y(x))$ in terms of its Taylor expansion about $(x, y(x))$, we obtain

$$f(x+h, y(x) + hf(x, y(x)))$$
$$= f(x, y(x)) + h\frac{\partial f}{\partial x}(x, y(x)) + hf(x, y(x))\frac{\partial f}{\partial y}(x, y(x)) + O(h^2).$$

Using $y'(x) = f(x, y(x))$ again, we get

$$\frac{df}{dx}(x, y(x)) = \frac{\partial f}{\partial x}(x, y(x)) + \frac{\partial f}{\partial y}(x, y(x))f(x, y(x)).$$

Hence

$$f(x+h, y(x) + hf(x, y(x))) = f(x, y(x)) + h\frac{df}{dx}(x, y(x)) + O(h^2)$$

or

$$0 = -f(x+h, y(x) + hf(x, y(x))) + f(x, y(x)) + h\frac{df}{dx}(x, y(x)) + O(h^2).$$
$$(7.6)$$

We multiply (7.6) by $-h/2$ and then add it to (7.5). We then have

$$y(x+h) = y(x) + \frac{h}{2}[f(x, y(x)) + f(x+h, y(x) + hf(x, y(x)))] + O(h^3).$$

This suggests the following method

$$\tilde{Y}_{i+1} = Y_i + hf(x_i, Y_i) \qquad (7.7)$$

$$Y_{i+1} = Y_i + \frac{h}{2}[f(x_i, Y_i) + f(x_{i+1}, \tilde{Y}_{i+1})]. \qquad (7.8)$$

7.3 Runge-Kutta methods

This method is called **Heun's method** or the second-order Runge-Kutta method. The local truncation error τ_i at x_i is defined by

$$\tau_i = \frac{1}{h}\left(y(x_{i+1}) - \left\{y(x_i) + \frac{h}{2}[f(x_i,y_i) + f(x_{i+1},y(x_i) + hf(x_i,y(x_i)))]\right\}\right),$$

and the global truncation error satisfies

$$\tau = \max |\tau_i| = O(h^2).$$

Thus the method is called a **second-order** method. The discretization error is

$$\max_i |y(x_i) - Y_i| = O(h^2).$$

See Exercise (7.4).

7.3.2 The fourth-order Runge-Kutta method

The fourth-order Runge-Kutta method is the most widely used numerical method, which is defined by

$$\begin{aligned}
k_1 &= hf(x_i, Y_i) \\
k_2 &= hf\left(x_i + \frac{h}{2}, Y_i + \frac{k_1}{2}\right) \\
k_3 &= hf\left(x_i + \frac{h}{2}, Y_i + \frac{k_2}{2}\right) \\
k_4 &= hf(x_i + h, Y_i + k_3) \\
Y_{i+1} &= Y_i + \frac{1}{6}(k_1 + 2k_2 + 2k_3 + k_4),
\end{aligned}$$

for $i = 0, 1, 2, \ldots, n-1$.

The local truncation error per step in this method is of order h^4, and the discretization error is

$$\max_i |y(x_i) - Y_i| = O(h^4).$$

See Exercise (7.5).

The reason for introducing the terminology k_1, k_2, k_3, k_4 into the method is to eliminate the terms with successive nesting in the second variable of

---

second-order method [二次法]　　successive nesting [遂次 k_i の定義式の第 2 変数のところに k_{i-1} が組み込まれていること]

$f(x, y)$. We leave the details as an exercise for the reader. The disadvantage of the fourth-order Runge-Kutta method is that it uses many more evaluations of f in each step as compared with the Euler method.

Example 7.2 Apply (a) the second-order Runge-Kutta method and (b) the fourth-order Runge-Kutta method to the following initial value problem

$$\frac{dy}{dx} = x + y, \qquad 0 \le x \le 1$$
$$y(0) = 0.$$

Use step-size $h = 0.1$.

Solution The mesh points are

$$x_0 = 0, \ x_1 = 0.1, \ x_2 = 0.2, \ \ldots, \ x_{10} = 1.$$

Let $f(x, y) = x + y$.

(a) Applying the second-order Runge-Kutta method (7.7) - (7.8) gives

$$\tilde{Y}_{i+1} = Y_i + 0.1(x_i + Y_i)$$
$$Y_{i+1} = Y_i + 0.05[x_i + Y_i + x_{i+1} + \tilde{Y}_{i+1}], \qquad i = 0, 1, \ldots, n-1.$$

We compute Y_i for $i = 1, 2, \ldots, n$ and obtain

$Y_1 = 0.005, \ Y_2 = 0.0105, \ Y_3 = 0.0326, \ Y_4 = 0.0675, \ Y_5 = 0.1166$

$Y_6 = 0.1814, \ Y_7 = 0.2634, \ Y_8 = 0.3646, \ Y_9 = 0.4868, \ Y_{10} = 0.6325.$

(b) Applying the fourth-order Runge-Kutta method gives

$$k_1 = 0.1(x_i + Y_i)$$
$$k_2 = 0.1(x_i + 0.05 + Y_i + 0.5k_1)$$
$$k_3 = 0.1(x_i + 0.05 + Y_i + 0.5k_2)$$
$$k_4 = 0.1(x_{i+1} + Y_i + k_3)$$
$$Y_{i+1} = Y_i + \frac{1}{6}(k_1 + 2k_2 + 2k_3 + k_4),$$

for $i = 0, 1, 2, \ldots, n-1$.

We compute Y_i for $i = 1, 2, \ldots, n$ and obtain

$Y_1 = 0.0052, \ Y_2 = 0.0214, \ Y_3 = 0.0499, \ Y_4 = 0.0918, \ Y_5 = 0.1487$

$Y_6 = 0.2221, \ Y_7 = 0.3138, \ Y_8 = 0.4255, \ Y_9 = 0.5596, \ Y_{10} = 0.7183.$

◇

7.4 Adams-Bashforth methods

The Euler method and Runge-Kutta methods are called **single-step** methods because the approximation Y_{i+1} of $y(x_{i+1})$ involves information from only one of the previous steps x_i and Y_i. Adams-Bashforth methods are multistep methods, which use the approximation from more than one previous steps to determine the approximation at the current point.

Integrating the initial value problem (7.1) - (7.2) from x_i to x_{i+1}, we have

$$\begin{aligned} y(x_{i+1}) - y(x_i) &= \int_{x_i}^{x_{i+1}} y'(x)dx \\ &= \int_{x_i}^{x_{i+1}} f(x,y(x))dx. \end{aligned}$$

We replace f by the mth interpolation polynomial $p_m(x)$ with the nodes

$$x_i, \ x_{i-1}, \ \ldots, \ x_{i-m},$$

and function values

$$f_i = f(x_i, y_i), \ f_{i-1} = f(x_{i-1}, y_{i-1}), \ \ldots,$$
$$f_{i-m} = f(x_{i-m}, y_{i-m}),$$

where $y_j = y(x_j)$, $j = i, i-1, \ldots, i-m$.

We use the Newton backward difference formula (5.9) to obtain

$$\begin{aligned} & p_m(x_i + \alpha h) \\ &= f_i + \alpha \nabla f_i + \frac{\alpha(\alpha+1)}{2} \nabla^2 f_i + \cdots \\ & \quad + \frac{\alpha(\alpha+1)\cdots(\alpha+m-1)}{m!} \nabla^m f_i \\ &= \sum_{j=0}^{m} (-1)^j \binom{-\alpha}{j} \nabla^j f_i \end{aligned}$$

where $\alpha = (x - x_i)/h$.

---

single-step method [一段法]   multistep method [多段法]

We integrate p_m with respect to x from x_i to x_{i+1}, so

$$\int_{x_i}^{x_{i+1}} p_m(x)dx$$

$$= h\int_0^1 p_m(x_i + \alpha h)d\alpha$$

$$= h\sum_{j=0}^m (-1)^j \left[\int_0^1 \binom{-\alpha}{j} d\alpha\right] \nabla^j f_i$$

$$= h\sum_{j=0}^m b_j f_{i-j}$$

where b_j are defined by the backward differences of f. See Section 5.4. This yields the $m+1$ step method

$$Y_{i+1} = Y_i + h\sum_{j=0}^m b_j f(x_{i-j}, Y_{i-j}),$$

which is called the $m+1$ **step Adams-Bashforth method**. The local truncation error τ_i at x_i is defined by

$$\tau_i = \frac{1}{h}[y(x_{i+1}) - y(x_i)] - \sum_{j=0}^m b_j f(x_{i-j}, y(x_{i-j})).$$

It can be shown that $\tau_i = O(h^{m+1})$ and

$$Y_i - y(x_i) = O(h^{m+1}).$$

In particular, we have

$$\int_0^1 p_0(x_i + \alpha h)h d\alpha = \int_0^1 f(x_i, y_i)h d\alpha = hf(x_i, y_i)$$

and

$$\int_0^1 p_1(x_i + \alpha h)h d\alpha = h\int_0^1 [f(x_i, y_i) + \alpha(f(x_i, y_i) - f(x_{i-1}, y_{i-1}))]d\alpha$$

$$= hf(x_i, y_i) + \frac{h}{2}(f(x_i, y_i) - f(x_{i-1}, y_{i-1}))$$

$$= \frac{h}{2}(3f(x_i, y_i) - f(x_{i-1}, y_{i-1})).$$

7.4 Adams-Bashforth methods

Similarly we can get the $m+1$ step Adams-Bashforth methods by integrating p_m for $m \geq 2$. The first four multistep Adams-Bashforth methods are given as follows.

One step Adams-Bashforth method ($m = 0$)

$$Y_{i+1} = Y_i + hf(x_i, Y_i), \qquad i = 0, 1, \ldots, n-1.$$

This is the Euler method.

Two step Adams-Bashforth method ($m = 1$)

$$Y_{i+1} = Y_i + \frac{h}{2}[3f(x_i, Y_i) - f(x_{i-1}, Y_{i-1})], \qquad i = 1, \ldots, n-1.$$

Three step Adams-Bashforth method ($m = 2$)

$$Y_{i+1} = Y_i + \frac{h}{12}[23f(x_i, Y_i) - 16f(x_{i-1}, Y_{i-1}) + 5f(x_{i-2}, Y_{i-2})],$$

$$i = 2, \ldots, n-1.$$

Four step Adams-Bashforth method ($m = 3$)

$$Y_{i+1} = Y_i + \frac{h}{24}[55f(x_i, Y_i) - 59f(x_{i-1}, Y_{i-1}) + 37f(x_{i-2}, Y_{i-2})$$

$$-9f(x_{i-3}, Y_{i-3})], \qquad i = 3, \ldots, n-1.$$

We need $m+1$ starting points for the $m+1$ step Adams-Bashforth method. Before applying the $m+1$ step method, we use a one step method, for instance, the fourth-order Runge-Kutta method, to get these $m+1$ starting points. For the $m+1$ step method if

$$|y(x_i) - Y_i| \leq \delta, \qquad i = 0, 1, \ldots, m$$

then there exist positive constants K_1, K_2 such that

$$|y(x_i) - Y_i| \leq K_1 \delta + K_2 h^m, \qquad i = m+1, \ldots.$$

Example 7.3 Apply the four step Adams-Bashforth method to the following initial value problem

$$\frac{dy}{dx} = x + y, \qquad 0 \leq x \leq 1$$
$$y(0) = 0$$

using $h = 0.1$, and the fourth-order Runge-Kutta method to calculate the starting points.

Solution The mesh points are
$$x_0 = 0, \ x_1 = 0.1, \ x_2 = 0.2, \ \ldots, \ x_{10} = 1.$$

Let $f(x, y) = x + y$. We need four starting points
$$(x_0, Y_0), \ (x_1, Y_1), \ (x_2, Y_2), \ (x_3, Y_3).$$

We calculate Y_1, Y_2 and Y_3 by the fourth-order Runge-Kutta method. Applying the four step Adams-Bashforth method gives
$$Y_{i+1} = Y_i + \frac{h}{24}[55(x_i + Y_i) - 59(x_{i-1} + Y_{i-1}) + 37(x_{i-2} + Y_{i-2})$$
$$-9(x_{i-3} + Y_{i-3})], \quad i = 3, 4, \ldots, n-1.$$

We compute Y_i for $i = 1, 2, \ldots, n$ and obtain

$Y_1 = 0.0052, \ Y_2 = 0.0214, \ Y_3 = 0.0499, \ Y_4 = 0.0918, \ Y_5 = 0.1487$

$Y_6 = 0.2221, \ Y_7 = 0.3138, \ Y_8 = 0.4255, \ Y_9 = 0.5596, Y_{10} = 0.7182.$ ◇

The exact solution of the initial value problem in Examples 7.1 – 7.3 is
$$y(x) = e^x - x - 1.$$

Thus the exact values of y (in 4 digits) at these mesh points are

$y(x_1) = 0.0052, \ y(x_2) = 0.0214, \ y(x_3) = 0.0499, \ y(x_4) = 0.0918,$

$y(x_5) = 0.1487, \ y(x_6) = 0.2221, \ y(x_7) = 0.3138, \ y(x_8) = 0.4255,$

$y(x_9) = 0.5596, \ y(x_{10}) = 0.7183.$

The errors of approximate values $\{Y_i\}$ at these mesh points obtained by the four methods discussed in this chapter are

Euler method $\quad \max_{1 \leq i \leq 10} |y(x_i) - Y_i| = 0.1245$

second-order Runge-Kutta method $\quad \max_{1 \leq i \leq 10} |y(x_i) - Y_i| = 0.0042$

fourth-order Runge-Kutta method $\quad \max_{1 \leq i \leq 10} |y(x_i) - Y_i| = 2.0843 \times 10^{-6}$

Four step Adams-Bashforth method $\quad \max_{1 \leq i \leq 10} |y(x_i) - Y_i| = 5.7389 \times 10^{-5}.$

The error strictly grows as the value of x_i increases, i.e.,
$$|y(x_i) - Y_i| > |y(x_{i-1}) - Y_{i-1}|, \quad i = 1, 2, \ldots, n.$$
in all of the four methods for this initial value problem. See Figure 7.2.

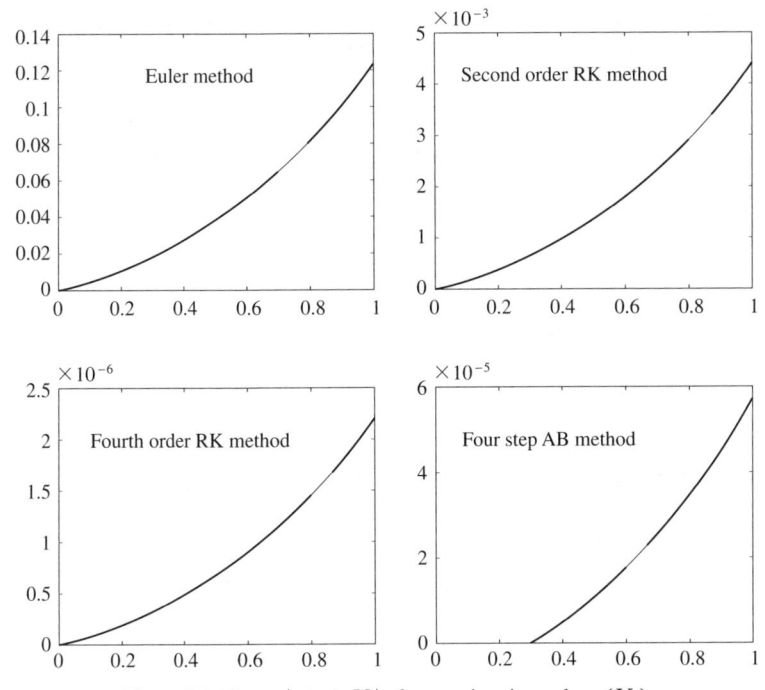

Figure 7.2: Errors $|y(x_i)-Y_i|$ of approximation values $\{Y_i\}$

Exercises

(7.1) Approximate the solution of the following initial value problem

$$\frac{dy}{dx} = y - x + 2$$
$$y(0) = 0$$

by using the four methods:

  (i) Euler method
  (ii) second-order Runge-Kutta method
  (iii) fourth-order Runge-Kutta method
  (vi) Four step Adams-Bashforth method

Write down 10 steps.

Compare the four methods.

(7.2) Divide the interval $[0, 1]$ into 4 and 8 subintervals of equal length. Approximate the solution of the initial value problem

$$\frac{dy}{dx} = -y + 2\cos x, \qquad 0 \leq x \leq 1$$
$$y(0) = 1$$

by using the three methods

(i) Euler method
(ii) second-order Runge-Kutta method
(iii) fourth-order Runge-Kutta method.

Compare the approximation obtained by each method with the true solution $y = \sin x + \cos x$.

(7.3) Divide the interval $[0, 1]$ into 8 subintervals of equal length. Use the four step Adams-Bashforth method to find an approximate solution of the initial value problem

$$\frac{dy}{dx} = y - 1, \qquad 0 \leq x \leq 1$$
$$y(0) = 2$$

(7.4) Divide the interval $[0, 1]$ into 4 subintervals of equal length. Use Heun's method to find an approximate solution of the initial value problem

$$\frac{dy}{dx} = y + \cos x - \sin x, \qquad 0 \leq x \leq 1$$
$$y(0) = 1$$

Estimate the error of the approximate solution.

(7.5) Divide the interval $[0, 1]$ into 2 subintervals of equal length. Use the fourth–order Runge-Kutta method to find an approximate solution of the initial value problem in Exercise (7.4). Estimate the error of the approximate solution.

Chapter 8.

Finite Difference Methods for Differential Equations

8.1 Two-point boundary value problems

In this section, we consider the linear two-point boundary value problem

$$-y'' + p(x)y' + q(x)y = r(x), \quad a \leq x \leq b \tag{8.1}$$
$$y(a) = \alpha, \quad y(b) = \beta, \tag{8.2}$$

where p, q and r are continuous functions on $[a, b]$.

It is known (Lees 1961) that if $q(x) \geq 0$ on $[a, b]$, then the problem (8.1)-(8.2) has a unique solution $y(x) \in C^2[a, b]$. Therefore, if $p, q, r \in C^2[a, b]$ and $q \geq 0$, then $y(x) \in C^4[a, b]$ (See [Keller 1968] or [Yamamoto 2003] for the proof in a special case $q(x) > 0$).

Finite difference methods for the boundary value problem replace the derivative y'' and y' in (8.1) by corresponding difference quotients at mesh points. This results a system of linear equations. By solving the system, we obtain approximation to the unique solution $y(x)$ at these mesh points.

---

finite difference method [有限差分法]   two-point boundary value problem [2 点境界値問題]

We choose an integer $n > 0$ and divide the interval $[a, b]$ into n equal subintervals, whose endpoints are the mesh points

$$x_i = a + ih, \quad i = 0, 1, \ldots, n, \quad \text{where} \quad h = \frac{b-a}{n}.$$

At the interior mesh points $x_i, i = 1, 2, \ldots, n-1$, we expand y in a Taylor polynomial about x_i,

$$y(x_{i+1}) = y(x_i) + hy'(x_i) + \frac{h^2}{2}y''(x_i) + \frac{h^3}{6}y^{(3)}(x_i) + \frac{h^4}{24}y^{(4)}(\xi_i) \quad (8.3)$$

and

$$y(x_{i-1}) = y(x_i) - hy'(x_i) + \frac{h^2}{2}y''(x_i) - \frac{h^3}{6}y^{(3)}(x_i) + \frac{h^4}{24}y^{(4)}(\eta_i), \quad (8.4)$$

where $\xi_i \in (x_i, x_{i+1})$ and $\eta_i \in (x_{i-1}, x_i)$. Adding (8.4) and (8.3), we obtain

$$y''(x_i) = \frac{1}{h^2}[y(x_{i+1}) - 2y(x_i) + y(x_{i-1})] - \frac{h^2}{24}(y^{(4)}(\xi_i) + y^{(4)}(\eta_i)). \quad (8.5)$$

Neglecting the last term gives the **centered difference formulate** for $y''(x_i)$

$$y''(x_i) \approx \frac{1}{h^2}[y(x_{i+1}) - 2y(x_i) + y(x_{i-1})].$$

We turn to the first derivative $y'(x_i)$. Subtracting (8.4) from (8.3), we obtain

$$y'(x_i) = \frac{1}{2h}[y(x_{i+1}) - y(x_{i-1})] - \frac{h^2}{6}y^{(3)}(x_i) - \frac{h^3}{48}(y^{(4)}(\xi_i) - y^{(4)}(\eta_i)). \tag{8.6}$$

Neglecting the terms in h^2 and h^3 gives the centered difference formula for $y'(x_i)$

$$y'(x_i) \approx \frac{1}{2h}[y(x_{i+1}) - y(x_{i-1})].$$

Substituting these centered difference formulas into the boundary value problem (8.1) - (8.2) and replaying $y(x_i)$ by their approximation Y_i, we obtain

$$-\frac{Y_{i+1} - 2Y_i + Y_{i-1}}{h^2} + p(x_i)\frac{Y_{i+1} - Y_{i-1}}{2h} + q(x_i)Y_i = r(x_i), \quad (8.7)$$

---

interior mesh point [内部分点]   centered difference formulate [中心差分公式]

8.1 Two-point boundary value problems

$$i = 1, 2, \ldots, n-1,$$

where $Y_0 = \alpha, Y_n = \beta$.

This is a **system of finite difference equations** corresponding to (8.1) - (8.2).

To simplify the notation, we set

$$p_i = p(x_i), \quad q_i = q(x_i), \quad r_i = r(x_i), \qquad i = 0, 1, \ldots, n.$$

We then rewrite (8.7) as

$$-\left(1 + \frac{h}{2}p_i\right)Y_{i-1} + (2 + h^2 q_i)Y_i - \left(1 - \frac{h}{2}p_i\right)Y_{i+1} = h^2 r_i.$$

Let

$$a_i = 2 + h^2 q_i, \qquad b_i = -1 - \frac{h}{2}p_i, \qquad c_i = -1 + \frac{h}{2}p_i, \qquad g_i = h^2 r_i.$$

The resulting system of equations is expressed with a tridiagonal $(n-1) \times (n-1)$ matrix A as

$$AY = g, \tag{8.8}$$

where

$$A = \begin{bmatrix} a_1 & c_1 & & & \\ b_2 & a_2 & c_2 & & \\ & \ddots & \ddots & \ddots & \\ & & \ddots & \ddots & c_{n-2} \\ & & & b_{n-1} & a_{n-1} \end{bmatrix}, \quad Y = \begin{bmatrix} Y_1 \\ Y_2 \\ \vdots \\ \vdots \\ Y_{n-1} \end{bmatrix},$$

$$g = \begin{bmatrix} g_1 + (1 + \frac{h}{2}p_1)\alpha \\ g_2 \\ \vdots \\ g_{n-2} \\ g_{n-1} + (1 - \frac{h}{2}p_{n-1})\beta \end{bmatrix}.$$

Solving the linear system by a numerical method discussed in Chapter 3, we obtain the approximate values $Y_i, i = 1, 2, \ldots, n-1$.

---

finite difference equations [有限差分方程式]

118   Chapter 8. Finite Difference Methods for Differential Equations

It is known that if $q_i \geq 0$ and h is sufficiently small, then the coefficient matrix A of the system of linear equations (8.8) is nonsingular and every element of A^{-1} is nonnegative. The local truncation error τ_i of the formula (8.7) at x_i is defined by

$$\begin{aligned}\tau_i &= -\frac{1}{h^2}[y(x_{i+1}) - 2y(x_i) + y(x_{i-1})] + p(x_i)\frac{y(x_{i+1}) - y(x_{i-1})}{2h} \\ &\quad + q(x_i)y(x_i) - r(x_i) \\ &= -\left\{\frac{1}{h^2}[y(x_{i+1}) - 2y(x_i) + y(x_{i-1})] - y''(x_i)\right\} \\ &\quad + p(x_i)\left\{\frac{y(x_{i+1}) - y(x_i)}{2h} - y'(x_i)\right\}.\end{aligned} \qquad (8.9)$$

We can write (8.9) as

$$A\begin{bmatrix} y(x_1) \\ \vdots \\ y(x_{n-1}) \end{bmatrix} = g + h^2\tau \quad \text{with} \quad \tau = \begin{bmatrix} \tau_1 \\ \vdots \\ \tau_{n-1} \end{bmatrix},$$

which implies, together with (8.8),

$$A\begin{bmatrix} y(x_1) - Y_1 \\ \vdots \\ y(x_{n-1}) - Y_{n-1} \end{bmatrix} = h^2\tau,$$

or

$$\begin{bmatrix} y(x_1) - Y_1 \\ \vdots \\ y(x_{n-1}) - Y_{n-1} \end{bmatrix} = h^2 A^{-1}\tau.$$

Furthermore, from (8.5) and (8.6), we have

$$\tau_i = -\frac{h^2}{24}(y^{(4)}(\xi_i) + y^{(4)}(\eta_i)) + p(x_i)\frac{h^2}{6}y^{(3)}(x_i) + \frac{p(x_i)}{48}h^3(y^{(4)}(\xi_i) - y^{(4)}(\eta_i)).$$

Let

$$P = \max_{x \in [a,b]} |p(x)| \quad \text{and} \quad M_j = \max_{x \in [a,b]} |y^{(j)}(x)|, \; j = 3, 4.$$

8.1 Two-point boundary value problems

Then

$$|\tau_i| \leq \frac{h^2}{12}M_4 + \frac{h^2 P}{6}M_3 + \frac{h^3 P}{24}M_4 = \frac{h^2}{12}[M_4 + 2PM_3 + \frac{h}{2}PM_4] \leq Ch^2,$$
$$i = 1, 2, \cdots, n-1,$$

where

$$C = \frac{1}{12}\left[M_4 + 2PM_3 + \frac{1}{2}PM_4\right],$$

which is independent of h.

Hence the error of the finite difference approximation can be estimated by

$$\begin{bmatrix} |y(x_1) - Y_1| \\ \vdots \\ |y(x_{n-1}) - y_{n-1}| \end{bmatrix} \leq h^2 A^{-1} \begin{bmatrix} |\tau_1| \\ \vdots \\ |\tau_{n-1}| \end{bmatrix} \leq Ch^2(h^2 A^{-1}) \begin{bmatrix} 1 \\ \vdots \\ 1 \end{bmatrix}, \quad (8.10)$$

where we understand that for two vectors $u = (u_i)$ and $v = (v_i) \in R^{n-1}$, $u \leq v$ means $u_i \leq v_i$, $i = 1, 2, \ldots, n-1$.

Furthermore, it can be shown that there exists a positive number M which is independent of h (and n) such that

$$h^2 A^{-1} \begin{bmatrix} 1 \\ \vdots \\ 1 \end{bmatrix} \leq M \begin{bmatrix} 1 \\ \vdots \\ 1 \end{bmatrix}.$$

To prove this, let $\varphi(x) \in C^2[a, b]$ be the unique solution of the boundary value problem

$$-y'' + p(x)y' + q(x)y = 2, \quad a \leq x \leq b$$
$$y(a) = y(b) = 0.$$

Then the above argument implies

$$\frac{1}{h^2} A \begin{bmatrix} \varphi(x_1) \\ \vdots \\ \varphi(x_{n-1}) \end{bmatrix} = \begin{bmatrix} 2 \\ \vdots \\ 2 \end{bmatrix} + \begin{bmatrix} \tilde{\tau}_1 \\ \vdots \\ \tilde{\tau}_{n-1} \end{bmatrix},$$

where

$$\tilde{\tau}_i = -\frac{1}{h^2}[\varphi(x_{i+1}) - 2\varphi(x_i) + \varphi(x_{i-1})] + p(x_i)\frac{\varphi(x_{i+1}) - \varphi(x_{i-1})}{2h}$$
$$+ q(x_i)\varphi(x_i) - r(x_i)$$
$$\to 0 \quad \text{as } h \to 0 \quad (\text{cf. [Yamamoto 2003(page 193)]}).$$

Hence, we have $|\tilde{\tau}_i| \leq 1$, $i = 1, 2, \ldots, n-1$ for sufficiently small $h > 0$ and

$$\frac{1}{h^2} A \begin{bmatrix} \varphi(x_1) \\ \vdots \\ \varphi(x_{n-1}) \end{bmatrix} \geq \begin{bmatrix} 1 \\ \vdots \\ 1 \end{bmatrix}.$$

Since each element of the matrix A^{-1} is nonnegative, we have

$$0 \leq h^2 A^{-1} \begin{bmatrix} 1 \\ \vdots \\ 1 \end{bmatrix} \leq \begin{bmatrix} \varphi(x_1) \\ \vdots \\ \varphi(x_{n-1}) \end{bmatrix} \leq M \begin{bmatrix} 1 \\ \vdots \\ 1 \end{bmatrix}$$

with $M = \max_{x \in [a,b]} \varphi(x) \geq 0$. We thus obtain from (8.10)

$$|y(x_i) - Y_i| \leq Ch^2 M, \quad i = 1, 2, \ldots, n-1.$$

Consequently, we have

$$\max_{1 \leq i \leq n-1} |y(x_i) - Y_i| = O(h^2).$$

Example 8.1 Use the finite difference method with $n = 10$ to approximate the solution to the linear boundary value problem

$$-y'' - y' + 2y = -2x - 1, \quad 0 \leq x \leq 1$$
$$y(0) = 0, \quad y(1) = e - 2.$$

Solution The mesh size is $h = 1/10 = 0.1$, and the mesh points are

$$x_i = ih, \quad i = 0, 1, \ldots, 10.$$

Since $p(x) = -1, q(x) = 2, r(x) = -2x - 1$, we have

$$a_i = 2(1+h^2) = 2.02, \quad b_i = -1 + \frac{h}{2} = -0.95, \quad c_i = -1 - \frac{h}{2} = -1.05,$$

8.2 Elliptic equations

$$g_i = -h^2(2x_i + 1) = -0.01 - 0.002 \times i.$$

Replacing the derivatives y'' and y' by difference quotients at these mesh points, we obtain the system of nine linear equations

$$\begin{bmatrix} 2.02 & -1.05 & & & & \\ -0.95 & 2.02 & -1.05 & & & \\ & \ddots & \ddots & \ddots & & \\ & & \ddots & \ddots & -1.05 \\ & & & -0.95 & 2.02 \end{bmatrix} \begin{bmatrix} Y_1 \\ Y_2 \\ \vdots \\ \vdots \\ Y_9 \end{bmatrix} = \begin{bmatrix} -0.012 \\ -0.014 \\ \vdots \\ \vdots \\ -0.028 + 1.05(e-2) \end{bmatrix}.$$

Solving the system by the Gaussian elimination method, we obtain

$$Y_1 = 0.0053, \ Y_2 = 0.0217, \ Y_3 = 0.0502, \ Y_4 = 0.0922, Y_5 = 0.1492$$
$$Y_6 = 0.2225, \ Y_7 = 0.3141, \ Y_8 = 0.4258, \ Y_9 = 0.5598.$$

The exact solution is

$$y(x) = e^x - x - 1.$$

For the error of Y_i at these mesh points, one has

$$\max_{1 \leq i \leq 9} |y(x_i) - Y_i| = |y(x_5) - Y_5| = 4.3549 \times 10^{-4}.$$

(cf. Examples 7.1-7.3). $\diamondsuit$

8.2 Elliptic equations

We consider the **Dirichlet problem**

$$-\Delta u(x,y) = f(x,y), \quad (x,y) \in \Omega \tag{8.11}$$
$$u(x,y) = g(x,y), \quad (x,y) \in \partial\Omega, \tag{8.12}$$

where
$$\Delta u(x,y) \equiv \frac{\partial^2 u}{\partial x^2}(x,y) + \frac{\partial^2 u}{\partial y^2}(x,y),$$

Ω is a bounded domain and $\partial\Omega$ denotes the boundary of Ω. It is known that if the boundary is piecewise smooth, $f \in C(\bar{\Omega}) \cap C^1(\Omega)$ and $g \in C(\bar{\Omega})$, then the problem has a unique solution $u \in C^2(\Omega) \cap C(\bar{\Omega})$. If the regularity of f and g will increase, then the regularity of u willll also increase. In the following, we assume that

$$\Omega = \{(x,y) \mid a < x < b, \ c < y < d\}$$

and $u \in C^4(\bar{\Omega})$.

The Dirichlet problem is one of most important problems in partial differential equations. The finite difference method stated in Section 8.1 is applicable to the two-dimensional problem (8.11)- (8.12).

The first step of the method is to choose integers n and m, and define mesh sizes
$$h = \frac{b-a}{n}, \qquad k = \frac{d-c}{m}.$$

Dividing the interval $[a,b]$ into n subintervals of equal width h and the interval $[c,d]$ into m subintervals of equal width k provides mesh points (x_i, y_j) where
$$x_i = a + ih, \quad i = 0, 1, \ldots, n$$
$$y_j = c + jk, \quad j = 0, 1, \ldots, m.$$

Drawing vertical and horizontal lines through these points with coordinates (x_i, y_j) generates a grid on the rectangle $[a,b] \times [c,d]$.

At each interior mesh point (x_i, y_j), $i = 1, 2, \ldots, n-1$, $j = 1, 2, \ldots, m-1$, we use the Taylor formula in the variable x about x_i to get

$$\begin{aligned}u(x_{i+1}, y_j) &= u(x_i, y_j) + h\frac{\partial u}{\partial x}(x_i, y_j) + \frac{h^2}{2}\frac{\partial^2 u}{\partial x^2}(x_i, y_j) \\ &\quad + \frac{h^3}{6}\frac{\partial^3 u}{\partial x^3}(x_i, y_j) + \frac{h^4}{24}\frac{\partial^4 u}{\partial x^4}(\xi_i, y_j)\end{aligned} \qquad (8.13)$$

and

$$\begin{aligned}u(x_{i-1}, y_j) &= u(x_i, y_j) - h\frac{\partial u}{\partial x}(x_i, y_j) + \frac{h^2}{2}\frac{\partial^2 u}{\partial x^2}(x_i, y_j) \\ &\quad - \frac{h^3}{6}\frac{\partial^3 u}{\partial x^3}(x_i, y_j) + \frac{h^4}{24}\frac{\partial^4 u}{\partial x^4}(\eta_i, y_j),\end{aligned} \qquad (8.14)$$

8.2 Elliptic equations

where $\xi_i \in (x_i, x_{i+1})$ and $\eta_i \in (x_{i-1}, x_i)$.

Adding (8.13) and (8.14) and neglecting terms in h^4, we obtain

$$\frac{\partial^2 u}{\partial x^2}(x_i, y_j) \approx \frac{1}{h^2}[u(x_{i+1}, y_j) - 2u(x_i, y_j) + u(x_{i-1}, y_j)]. \qquad (8.15)$$

Similarly, we use the Taylor formula in the variable y about y_j and obtain

$$\frac{\partial^2 u}{\partial y^2}(x_i, y_j) \approx \frac{1}{k^2}[u(x_i, y_{j+1}) - 2u(x_i, y_j) + u(x_i, y_{j-1})]. \qquad (8.16)$$

Let $U_{i,j}$ be the approximation to $u(x_i, y_j)$ and

$$f_{i,j} = f(x_i, y_j), \qquad i = 1, 2, \ldots, n-1, \; j = 1, 2, \ldots, m-1$$

at interior mesh points and

$$U_{0,j} = g(x_0, y_j), \quad U_{n,j} = g(x_n, y_j), \qquad j = 1, \ldots, m-1$$
$$U_{i,0} = g(x_i, y_0), \quad U_{i,m} = g(x_i, y_m), \qquad i = 1, 2, \ldots, n-1$$

at boundary mesh points. See Figure 8.1.

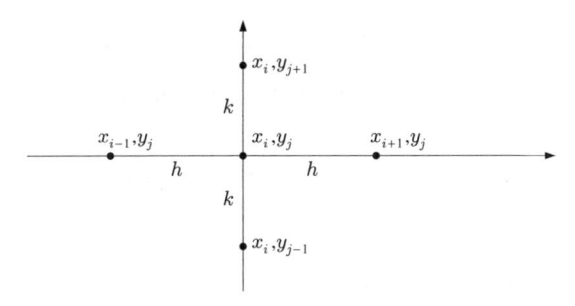

Figure 8.1: Points in (8.15) and (8.16)

Substituting

$$\frac{\partial^2 u}{\partial x^2} \approx \frac{U_{i-1,j} - 2U_{i,j} + U_{i+1,j}}{h^2}$$

and

$$\frac{\partial^2 u}{\partial y^2} \approx \frac{U_{i,j-1} - 2U_{i,j} + U_{i,j+1}}{k^2}$$

into (8.11), we obtain

$$\frac{-U_{i-1,j} + 2U_{i,j} - U_{i+1,j}}{h^2} + \frac{-U_{i,j-1} + 2U_{i,j} - U_{i,j+1}}{k^2} = f_{i,j} \qquad (8.17)$$

for $i = 1, 2, \ldots, n-1$ and $j = 1, 2, \ldots, m-1$. This is called the **centered difference formula** or **five-point formula** for $-\triangle u(x_i, y_j)$. The local truncation error of the formula (8.17) is $O(h^2 + k^2)$, where the local truncation error τ_{ij} at (x_i, y_j) is defined by

$$\begin{aligned}\tau_{ij} &= \frac{-u(x_{i-1}, y_j) + 2u(x_i, y_j) - u(x_{i+1}, y_j)}{h^2} \\ &\quad + \frac{-u(x_i, y_{j-1}) + 2u(x_i, y_j) - u(x_i, y_{j+1})}{k^2} - f(x_i, y_j) \\ &= \frac{-u(x_{i-1}, y_j) + 2u(x_i, y_j) - u(x_{i+1}, y_j)}{h^2} \\ &\quad + \frac{-u(x_i, y_{j-1}) + 2u(x_i, y_j) - u(x_i, y_{j+1})}{k^2} + \triangle u(x_i, y_j).\end{aligned}$$

For the case $a = c$ and $b = d$, if we choose $h = k$ (hence $m = n$) in (8.17) and order the approximations $U_{i,j}$ of the true values $u(x_i, y_j)$ at the interior mesh points in $(a,b) \times (c,d)$ as

$$U = [U_{1,1}, U_{2,1}, \ldots, U_{n-1,1}, \ldots, U_{1,m-1}, U_{2,m-1}, \ldots, U_{n-1,m-1}]^T,$$

then we obtain a system of linear equations

$$AU = g + h^2 f, \tag{8.18}$$

where g and f are $(m-1)(n-1) = (n-1)^2$ dimensional vectors defined by

$$g = \begin{bmatrix} [g_{1,0} + g_{0,1}, g_{2,0}, \ldots, g_{n-2,0}, g_{n-1,0} + g_{n,1}]^T \\ [g_{0,2}, 0, \ldots, 0, g_{n,2}]^T \\ \cdots \\ [g_{0,m-2}, 0, \ldots, 0 g_{n,m-2}]^T \\ [g_{0,m-1} + g_{1,m}, g_{2,m}, \ldots, g_{n-2,m}, g_{n-1,m} + g_{n,m-1}]^T \end{bmatrix}$$

and

$$f = [f_{1,1}, \ldots, f_{n-1,1}, \ldots, f_{1,m-1}, f_{2,m-1}, \ldots, f_{n-1,m-1}]^T,$$

$g_{i,j} = g(x_i, y_j)$, and A is a block tri-diagonal $(m-1)(n-1) \times (m-1)(n-1)$ matrix

$$A = \begin{bmatrix} H & -I & & & \\ -I & H & -I & & \\ & \ddots & \ddots & \ddots & \\ & & \ddots & \ddots & -I \\ & & & -I & H \end{bmatrix}$$

8.2 Elliptic equations

where

$$H = \begin{bmatrix} 4 & -1 & & & \\ -1 & 4 & -1 & & \\ & \ddots & \ddots & \ddots & \\ & & \ddots & \ddots & -1 \\ & & & -1 & 4 \end{bmatrix} \quad ((n-1) \times (n-1))$$

and I is the $(n-1) \times (n-1)$ identity matrix. Then we have

$$\frac{1}{h^2} A(\boldsymbol{u} - U) = \tau,$$

where $\boldsymbol{u}$ is the vector which is obtained by replacing each element $U_{i,j}$ of U by $u(x_i, y_j)$ and $\tau = [\tau_{11}, \tau_{21}, \ldots, \tau_{n-1.m-1}]^T$, since

$$A\boldsymbol{u} = g + h^2 f + h^2 \tau.$$

It can be shown that A is nonsingular, $\|h^2 A^{-1}\|_\infty$ is bounded and

$$\|\boldsymbol{u} - U\|_\infty = \|h^2 A^{-1} \tau\|_\infty \leq \|h^2 A^{-1}\|_\infty \|\tau\|_\infty = O(h^2).$$

Example 8.2 Apply the finite difference method to solve the Dirichlet problem

$$-\left[\frac{\partial^2 u}{\partial x^2}(x,y) + \frac{\partial^2 u}{\partial y^2}(x,y)\right] = 2\pi^2 \sin \pi x \sin \pi y, \quad (x,y) \in (0,1) \times (0,1)$$
$$u(0,y) = u(1,y) = u(x,0) = u(x,1) = 0$$

with $n = 4$ and $m = 4$.

Solution The mesh size is $h = k = \dfrac{1}{4} = 0.25$. The mesh points are

$$(x_i, y_j) = (ih, jh), \quad i,j = 0,1,2,3,4,$$

The number of the interior mesh points is 9. Applying the finite difference formula (8.17) to this problem gives a 9 by 9 linear system (8.18)

where

$$A = \begin{bmatrix} 4 & -1 & 0 & -1 & 0 & 0 & 0 & 0 & 0 \\ -1 & 4 & -1 & 0 & -1 & 0 & 0 & 0 & 0 \\ 0 & -1 & 4 & 0 & 0 & -1 & 0 & 0 & 0 \\ -1 & 0 & 0 & 4 & -1 & 0 & -1 & 0 & 0 \\ 0 & -1 & 0 & -1 & 4 & -1 & 0 & -1 & 0 \\ 0 & 0 & -1 & 0 & -1 & 4 & 0 & 0 & -1 \\ 0 & 0 & 0 & -1 & 0 & 0 & 4 & -1 & 0 \\ 0 & 0 & 0 & 0 & -1 & 0 & -1 & 4 & -1 \\ 0 & 0 & 0 & 0 & 0 & -1 & 0 & -1 & 4 \end{bmatrix}$$

$$U = \begin{bmatrix} U_{1,1} \\ U_{2,1} \\ U_{3,1} \\ U_{1,2} \\ U_{2,2} \\ U_{3,2} \\ U_{1,3} \\ U_{2,3} \\ U_{3,3} \end{bmatrix} \quad \text{and} \quad f = \begin{bmatrix} f_{1,1} \\ f_{2,1} \\ f_{3,1} \\ f_{1,2} \\ f_{2,2} \\ f_{3,2} \\ f_{1,3} \\ f_{2,3} \\ f_{3,3} \end{bmatrix} = \begin{bmatrix} 2\pi^2 \sin \pi x_1 \sin \pi y_1 \\ 2\pi^2 \sin \pi x_2 \sin \pi y_1 \\ 2\pi^2 \sin \pi x_3 \sin \pi y_1 \\ 2\pi^2 \sin \pi x_1 \sin \pi y_2 \\ 2\pi^2 \sin \pi x_2 \sin \pi y_2 \\ 2\pi^2 \sin \pi x_3 \sin \pi y_2 \\ 2\pi^2 \sin \pi x_1 \sin \pi y_3 \\ 2\pi^2 \sin \pi x_2 \sin \pi y_3 \\ 2\pi^2 \sin \pi x_3 \sin \pi y_3 \end{bmatrix},$$

and the boundary condition yields

$$g = [g_{0,1} + g_{1,0}, g_{2,0}, g_{3,0} + g_{4,1}, g_{0,2}, 0, g_{4,2}, g_{0,3} + g_{1,4}, g_{2,4}, g_{3,4} + g_{4,3}]^T = 0.$$

See Figure 8.2.
Solving this linear system

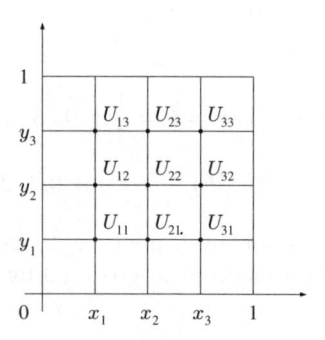

Figure 8.2: Finite difference mesh in Example 8.2

8.3 Parabolic equations

$$AU = h^2 f,$$

we obtain approximate values

$U_{1,1} = 0.5265$, $U_{2,1} = 0.7446$, $U_{3,1} = 0.5265$, $U_{1,2} = 0.7446$,

$U_{2,2} = 1.0530$, $U_{3,2} = 0.7446$, $U_{1,3} = 0.5265$, $U_{2,3} = 0.7446$, $U_{3,3} = 0.5265$

See Figure 8.3.

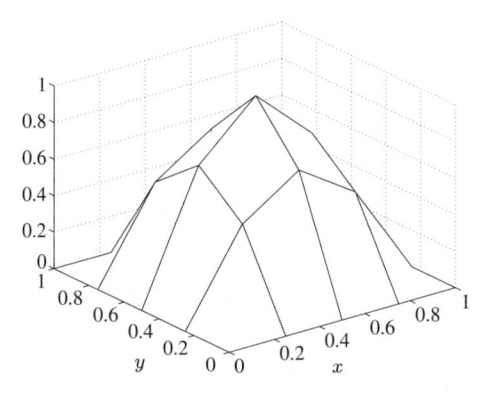

Figure 8.3: Numerical solution of Example 8.2

The exact solution is

$$u = \sin \pi x \sin \pi y.$$

The error of U has

$$\max_{\substack{1 \leq i \leq 3 \\ 1 \leq j \leq 3}} |u(x_i, y_j) - U_{i,j}| = |u(x_2, y_2) - U_{2,2}| = 0.0530. \qquad \diamond$$

8.3 Parabolic equations

The parabolic partial differential equation we consider in this section is the one-dimensional heat equation

$$\frac{\partial u}{\partial t}(x,t) = \frac{\partial^2 u}{\partial x^2}(x,t), \quad 0 < x < 1, \quad t > 0 \qquad (8.19)$$

$$u(x,0) = f(x), \quad 0 \leq x \leq 1, \text{initial condition} \qquad (8.20)$$

$$u(0,t) = u(1,t) = 0, \quad t \geq 0, \text{boundary conditions}. \qquad (8.21)$$

Chapter 8. Finite Difference Methods for Differential Equations

To find approximate values of $u(x,t)$ for $0 < x < 1$, $0 < t \leq T$, we first choose two mesh sizes $h = 1/n$ and $k = T/m$ in x and t directions, respectively, and set the mesh points

$$(x_i, t_j) = (ih, jk), \qquad i = 0, 1, \ldots, n, \ j = 0, 1, \ldots, m.$$

Applying the centered difference formula for $\partial^2 u/\partial x^2$ and the forward difference formula for $\partial u/\partial t$ to (8.19), we obtain

$$\frac{1}{k}(U_{i,j+1} - U_{i,j}) = \frac{1}{h^2}(U_{i+1,j} - 2U_{i,j} + U_{i-1,j}), \qquad (8.22)$$

where $U_{i,j}$ denote approximations of $u(x_i, t_j)$, $0 \leq i \leq n, 0 \leq j \leq m$ (See Figure 8.4).

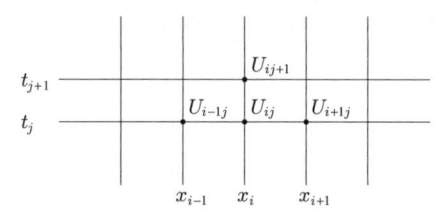

Figure 8.4: Points in (8.22)

We rewrite (8.22) as

$$U_{i,j+1} = U_{i,j} + \frac{k}{h^2}(U_{i+1,j} - 2U_{i,j} + U_{i-1,j}). \qquad (8.23)$$

By (8.23), $\{U_{i,j+1}\}_{i=1}^{n-1}$ is determined from $\{U_{i,j}\}_{i=1}^{n-1}$, for $j = 1, 2, \ldots, m$. This method is called **direct method**.

At the first step of this method, we compute

$$U_{i,1} = U_{i,0} + \frac{k}{h^2}(U_{i+1,0} - 2U_{i,0} + U_{i-1,0}),$$

where $U_{i,1}$ corresponds to time row t_1, in terms of the three other U that correspond to time row t_0 which are known by the initial condition (8.20). After computing $U_{1,1}, U_{2,1}, \ldots, U_{n-1,1}$, we compute $U_{1,2}, U_{2,2}, \ldots, U_{n-1,2}$ in a similar manner, etc.

---

direct method [直接法]

8.3 Parabolic equations

Let

$$U^{(j)} = \begin{bmatrix} U_{1,j} \\ U_{2,j} \\ \vdots \\ U_{n-1,j} \end{bmatrix} \quad \text{and} \quad A = \begin{bmatrix} 1-2\mu & \mu & & & \\ \mu & 1-2\mu & \mu & & \\ & \ddots & \ddots & \ddots & \\ & & \ddots & \ddots & \mu \\ & & & \mu & 1-2\mu \end{bmatrix},$$

where $\mu = k/h^2$. Then the direct method (8.23) can be written in the form

$$U^{(j+1)} = AU^{(j)}.$$

The method is said to be stable if for any $T > 0$, there exists a constant $K > 0$, independent of h and k such that, for any initial data $U^{(0)}$, the sequence $\{U^{(j)}\}$ satisfies

$$\|U^{(j)}\|_2 = \|A^j U^{(0)}\|_2 \leq K \|U^{(0)}\|_2, \quad j = 0, 1, 2, \cdots, m,$$

which is equivalent to

$$\|A^j\|_2 \leq K < +\infty, \quad j = 0, 1, 2, \cdots.$$

Since A is symmetric, we have $\|A^j\|_2 = \rho(A^j)$ so that the stability condition $\|A^j\|_2 \leq K < +\infty$ reduces to $\rho(A) = \max_i |\lambda_i| \leq 1$. This means that even if the initial value $U^{(0)}$ is replaced by erroneous value $\tilde{U}^{(0)} = U^{(0)} + \varepsilon$, the method (8.23) yields a stable sequence $\{\tilde{U}^{(j)}\}$ which satisfies

$$\|\tilde{U}^{(j)} - U^{(j)}\|_2 \leq \|\varepsilon\|_2,$$

since

$$\begin{aligned} \|\tilde{U}^{(j)} - U^{(j)}\|_2 &= \|A^j(\tilde{U}^{(0)} - U^{(0)})\|_2 \\ &= \|A^j \varepsilon\|_2 \\ &\leq \|A\|_2^j \|\varepsilon\|_2 \\ &= \rho(A)^j \|\varepsilon\|_2 \\ &\leq \|\varepsilon\|_2. \end{aligned}$$

It is known that the eigenvalues of the tridiagonal matrix A is

$$\lambda_i = 1 - 4\mu \sin^2 \frac{i\pi}{2n}, \quad i = 1, 2, \ldots, n-1.$$

See [Yamamoto, 2003].

Hence, to ensure the stability of the method, we must have for any n

$$-1 \le 1 - 4\mu \sin^2 \frac{i\pi}{2n} \le 1, \quad 1 \le i \le n-1,$$

which implies

$$\mu \sin^2 \frac{i\pi}{2n} \le \frac{1}{2}, \quad 1 \le i \le n-1.$$

To ensure that this inequality holds for all n, we must have

$$\mu = \frac{k}{h^2} \le \frac{1}{2}. \tag{8.24}$$

This means that we should not move too fast in the t-direction.

Furthermore, under the condition (8.24), the approximation $\{U_{ij}\}$ converges to the exact solution $\{u(x_i, t_j)\}$ as $h, k \to 0$. The proof can be found in [Yamamoto 2003].

Example 8.3 Solve the heat equation

$$\frac{\partial u}{\partial t}(x,t) = \frac{\partial^2 u}{\partial x^2}(x,t), \quad 0 < x < 1, \quad t > 0$$
$$u(x,0) = \sin \pi x, \quad 0 < x < 1$$
$$u(0,t) = u(1,t) = 0, \quad t \ge 0.$$

Solution To ensure the convergence of the direct method, we choose $h = 1/4$ and $k = 1/32$, and thus (8.24) holds. We compute $U_{i,j}$ at (x_i, t_j), where

$$x_i = ih, \ i = 1,2,3, \ \text{ and } \ t_j = jk, j \ge 1.$$

The initial condition gives

$$U_{i,0} = \sin ih\pi, \quad i = 0,1,2,3,4$$

and the boundary condition gives

$$U_{0,j} = U_{n,j} = 0, \quad j \ge 0.$$

We give results computed by (8.23) for $j = 1, 2, 3, 4$.

$$[U_{1,1} \ U_{2,1} \ U_{3,1}] = [0.5000 \quad 0.7071 \quad 0.5000]$$
$$[U_{1,2} \ U_{2,2} \ U_{3,2}] = [0.3536 \quad 0.5000 \quad 0.3536]$$
$$[U_{1,3} \ U_{2,3} \ U_{3,3}] = [0.2500 \quad 0.3536 \quad 0.2500]$$
$$[U_{1,4} \ U_{2,4} \ U_{3,4}] = [0.1768 \quad 0.2500 \quad 0.1768]$$
$$[U_{1,5} \ U_{2,5} \ U_{3,5}] = [0.1250 \quad 0.1768 \quad 0.1250]$$

8.4 Hyperbolic equations

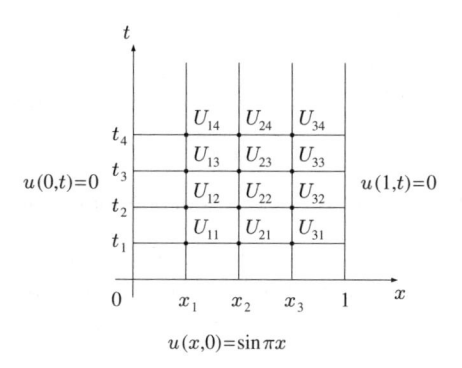

Figure 8.5: Finite difference grid in Example 8.3

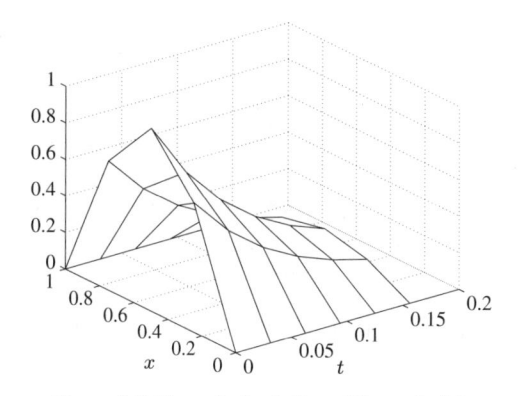

Figure 8.6: Numerical solution of Example 8.3

The exact solution is

$$u(x,t) = e^{-\pi^2 t} \sin \pi x.$$ ◇

8.4 Hyperbolic equations

In this section, we consider the numerical solution to the wave equation, a typical example of a hyperbolic equation, which is given by

Chapter 8. Finite Difference Methods for Differential Equations

$$\frac{\partial^2 u}{\partial t^2}(x,t) = \alpha^2 \frac{\partial^2 u}{\partial x^2}(x,t), \qquad 0 \le x \le 1, \quad t \ge 0 \qquad (8.25)$$
$$u(x,0) = f(x), \qquad 0 \le x \le 1, \text{initial displacement} \quad (8.26)$$
$$\frac{\partial u}{\partial t}(x,0) = g(x), \qquad 0 \le x \le 1, \text{initial velocity} \qquad (8.27)$$
$$u(0,t) = u(1,t) = 0, \qquad t > 0, \text{boundary conditions}, \qquad (8.28)$$

where α is a constant. See [Farlow 1982] for derivation of such a problem.

The first step is similar to the finite difference method for the heat equation. We select two mesh sizes $h = 1/n$ and $k = T/m$, and set the mesh points

$$(x_i, t_j) = (ih, jk), \qquad i = 0, 1, \ldots n, \quad j = 0, 1, \ldots m.$$

Applying the centered difference formula for $\partial^2 u/\partial x^2$ and $\partial^2 u/\partial t^2$ to (8.25), we obtain

$$\frac{1}{k^2}(U_{i,j+1} - 2U_{i,j} + U_{i,j-1}) = \frac{\alpha^2}{h^2}(U_{i+1,j} - 2U_{i,j} + U_{i-1,j}), \qquad (8.29)$$

with the truncation error $O(h^2 + k^2)$. If we set $\mu = \frac{\alpha k}{h}$, then we have

$$U_{i,j+1} = 2(1-\mu^2)U_{i,j} + \mu^2(U_{i-1,j} + U_{i+1,j}) - U_{i,j-1}. \qquad (8.30)$$

This method is a **direct method**. See Figure 8.7. We have

$$\begin{bmatrix} U_{1,j+1} \\ U_{2,j+1} \\ \vdots \\ U_{n-1,j+1} \end{bmatrix}$$

$$= \begin{bmatrix} 2(1-\mu^2) & \mu^2 & & \\ \mu^2 & \ddots & \ddots & \\ & \ddots & \ddots & \mu^2 \\ & & \mu^2 & 2(1-\mu^2) \end{bmatrix} \begin{bmatrix} U_{1,j} \\ U_{2,j} \\ \vdots \\ U_{n-1,j} \end{bmatrix} - \begin{bmatrix} U_{1,j-1} \\ U_{2,j-1} \\ \vdots \\ U_{n-1,j-1} \end{bmatrix}.$$

The difference from the direct method for the heat equation is that (8.30) involves 3 time steps $j-1$, j, $j+1$, whereas the formulas for the heat equation involved only 2 time steps. The method (8.30) has a minor problem at the starting values for $j = 0$. Now we show how to get started by using the two initial conditions (8.26) and (8.27).

8.4 Hyperbolic equations

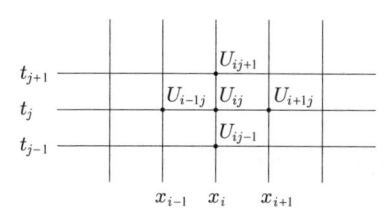

Figure 8.7: Points in (8.30)

From (8.26), we have

$$U_{i,0} = u(x_i, 0) = f(x_i), \quad i = 0, 1, \ldots, n.$$

From (8.27) we derive the difference formula

$$\frac{1}{2k}(U_{i,1} - U_{i,-1}) = g_i.$$

Thus

$$U_{i,1} = U_{i,-1} + 2kg_i. \tag{8.31}$$

For $t = 0$, i.e., $j = 0$, (8.30) is

$$U_{i,1} = 2(1-\mu^2)U_{i,0} + \mu^2(U_{i-1,0} + U_{i+1,0}) - U_{i,-1}. \tag{8.32}$$

Adding (8.32) to (8.31) and dividing by 2, we obtain

$$U_{i,1} = (1-\mu^2)U_{i,0} + \frac{\mu^2}{2}(U_{i-1,0} + U_{i+1,0}) + kg_i \tag{8.33}$$

which expresses $U_{i,1}$ in terms of the initial data.

We summarize the direct method for the wave equation as follows:

For $j = 0$
$$U_{i,0} = f(x_i)$$

For $j = 1$
$$U_{i,1} = (1-\mu^2)U_{i,0} + \frac{\mu^2}{2}(U_{i-1,0} + U_{i+1,0}) + kg_i \tag{8.34}$$

For $j \geq 2$
$$U_{i,j} = 2(1-\mu^2)U_{i,j-1} + \mu^2(U_{i-1,j-1} + U_{i+1,j-1}) - U_{i,j-2}.$$
$$i = 1, 2, \ldots, n. \tag{8.35}$$

Chapter 8. Finite Difference Methods for Differential Equations

To ensure the convergence of this method, we must have

$$\mu = \frac{\alpha k}{h} \leq 1.$$

See [Isaacson-Keller, 1966].

Example 8.4 Solve the wave equation

$$\frac{\partial^2 u}{\partial t^2}(x,t) = 4\frac{\partial^2 u}{\partial x^2}(x,t), \qquad 0 \leq x \leq 1, \quad t \geq 0$$
$$u(x,0) = \sin \pi x, \qquad 0 \leq x \leq 1$$
$$\frac{\partial u}{\partial t}(x,0) = 0, \qquad 0 \leq x \leq 1$$
$$u(0,t) = u(1,t) = 0, \qquad t > 0.$$

Solution We choose

$$h = \frac{1}{4}, \qquad k = \frac{h}{\alpha} = \frac{1}{8}.$$

Then the mesh points are

$$x_i = ih, \quad i = 0,1,2,3,4, \qquad t_j = jk, \quad j \geq 0.$$

The first initial condition gives

$$U_{i,0} = \sin ih\pi$$

and the boundary condition gives

$$U_{0,j} = U_{4,j} = 0, \quad j \geq 0.$$

In this example, $g_i = 0$.

Since $\mu = \alpha k/h = 1$, $U_{i,j-1}$ drops out in (8.34) and (8.35) and we have

$$U_{i,j} = \begin{cases} \frac{1}{2}(U_{i-1,j-1} + U_{i+1,j-1}), & j = 1 \\ U_{i-1,j-1} + U_{i+1,j-1} - U_{i,j-2}, & j \geq 2. \end{cases}$$

We use the above formula to obtain U_{ij}.

We give results for $j = 1, 2, 3$. The numerical solution $\{U_{ij}\}$, $1 \leq i \leq 4, 1 \leq j \leq 8$ are shown in Figures 8.8, 8.9.

$$U_{1,1} = \frac{1}{2}(U_{0,0} + U_{2,0}) = \frac{1}{2}\sin 2h\pi = 0.5000$$
$$U_{2,1} = \frac{1}{2}(U_{1,0} + U_{3,0}) = \frac{1}{2}(\sin h\pi + \sin 3h\pi) = 0.7071$$
$$U_{3,1} = \frac{1}{2}(U_{2,0} + U_{4,0}) = \frac{1}{2}\sin 2h\pi = 0.5000.$$

8.4 Hyperbolic equations

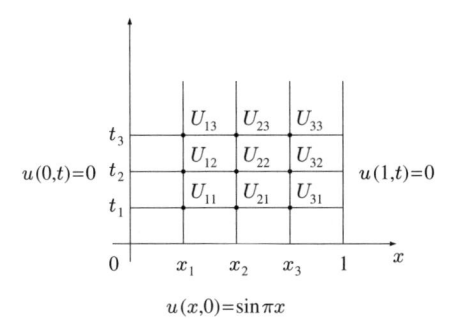

Figure 8.8: Finite difference mesh in Example 8.4

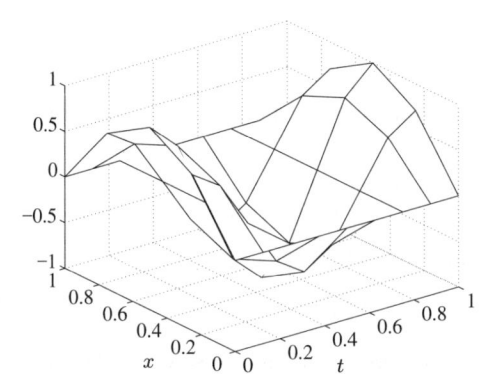

Figure 8.9: Numerical solution of Example 8.4

$$U_{12} = U_{0,1} + U_{2,1} - U_{1,0} = 0$$
$$U_{22} = U_{1,1} + U_{3,1} - U_{2,0} = 0$$
$$U_{32} = U_{2,1} + U_{4,1} - U_{3,0} = 0$$
$$U_{13} = U_{0,2} + U_{2,2} - U_{1,1} = -0.5000$$
$$U_{23} = U_{1,2} + U_{3,2} - U_{2,1} = -0.7071$$
$$U_{33} = U_{2,2} + U_{4,2} - U_{3,1} = -0.5000.$$

The exact solution is

$$u(x,t) = \sin \pi x \cos 2\pi t.$$

Exercises

(8.1) Use the finite difference method with $n=4$ to solve the boundary value problem
$$y'' = -y' + 2y - 2x - 1, \quad 0 \le x \le 1$$
$$y(0) = 2, \quad y(1) = e + 2.$$

(8.2) Use the finite difference method with $n = m = 4$ to solve the Dirichlet problem
$$\frac{\partial^2 u}{\partial x^2} + \frac{\partial^2 u}{\partial y^2} = 2(x^2 + y^2), \quad (x,y) \in (0,1) \times (0,1)$$
$$u(0,y) = u(x,0) = 0, u(1,y) = y^2, u(x,1) = x^2.$$

(8.3) Use the finite difference method with $n = m = 4$ to solve the Dirichlet problem
$$\frac{\partial^2 u}{\partial x^2} + \frac{\partial^2 u}{\partial y^2} = 2(x^2 + y^2), \quad (x,y) \in (0,1) \times (0,1)$$
$$u(x,0) = 0, u(x,1) = x^2, \quad 0 \le x \le 1$$
$$u(0,y) = \sin \pi, u(1,y) = 2e^\pi \sin \pi y + y^2, \quad 0 \le y \le 1.$$

(8.4) Use the finite difference method to solve the Dirichlet problem
$$\frac{\partial^2 u}{\partial x^2} + \frac{\partial^2 u}{\partial y^2} = -\pi^2 (\sin \pi x + \cos \pi y), \quad (x,y) \in (0,1) \times (0,1)$$
$$u(0,y) = u(x,0) = u(1,y) = u(x,1) = 1$$

with the mesh points
$$x_0 = y_0 = 0, \; x_1 = y_1 = \frac{1}{3}, \; x_2 = y_2 = \frac{2}{3}, \; x_3 = y_3 = 1.$$

(8.5) Suppose that $u(x,t)$ is a C^4 function with respect to x and t. Give the truncation error in the finite difference approximation
$$\frac{u(x, t+k) - u(x,t)}{k} = \frac{u(x+h, t) - 2u(x,t) + u(x-h, t)}{h^2}$$
for
$$\frac{\partial u}{\partial t} = \frac{\partial^2 u}{\partial x^2}.$$

Exercises

(8.6) Use the finite difference method with $h = 1/4$ and $k = 1/32$ to solve the heat equation

$$\frac{\partial u}{\partial t} = \frac{\partial^2 u}{\partial x^2}, \qquad 0 < x < 2, t > 0$$
$$u(x, 0) = \sin \frac{\pi}{2} x, \qquad 0 \leq x \leq 2$$
$$u(0, t) = u(2, t) = 0, \qquad t > 0.$$

(8.7) Use the finite difference method with $h = 1/4$ and $k = 1/4$ to solve the wave equation

$$\frac{\partial^2 u}{\partial t^2} = \frac{\partial^2 u}{\partial x^2}, \qquad 0 < x < 1, t > 0$$
$$u(x, 0) = \sin \pi x, \qquad 0 \leq x \leq 1$$
$$\frac{\partial u}{\partial t}(x, 0) = 0, \qquad 0 \leq x \leq 1$$
$$u(0, t) = u(1, t) = 0, \qquad t > 0.$$

付録 A　数学式の読み方

Addition 加法

$a + b = c$　　　　a plus b equals c

$967 + 45 = 1012$　　nine hundred sixty-seven plus forty-five equals one thousand and twelve

Subtraction 減法

$a - b = c$　　　　a minus b equals c

$978 - 562 = 416$　　nine hundred seventy-eight minus five hundred sixty-two equals four hundred and sixteen

Multiplication 乗法

$a \times b = c$　　　a times b equals c

$16.25 \times 5 = 81.25$　sixteen point two five times five equals eighty-one point twenty five

Division 除法

$a \div b = c$　　　a divided by b equals c

$33 \div 3 = 11$　　thirty three divided by three equals eleven

Fractions 分数

$\dfrac{1}{4}$　　　　one fourth, a quarter, one over four

$2\dfrac{3}{10}$　　　two and three-tenths

$\dfrac{1}{2} + \dfrac{1}{3} = \dfrac{5}{6}$　one half plus one third equals five-sixths

one over two plus one over three equals five over six

Powers べき乗

$4^2 = 16$　　　four squared equals sixteen

$5^3 = 125$　　five cubed equals one hundred twenty five

x^4　　　　　x to the fourth power

the fourth power of x

付録 A　　数学式の読み方

	x fourth
a^{-n}	a to the minus nth power
	the minus nth power of a

Roots 根

$\sqrt{4} = 2$	the square root of 4 equals 2
$\sqrt[3]{8} = 2$	the cube root of 8 equals 2
$\sqrt[n]{x} = y$	the nth root of x equals y
$\sqrt{b^2 - 4ac}$	the square root of the quantity b squared minus $4ac$
$a^2 - b^2 = (a+b)(a-b)$	a squared minus b squared equals the quantity a plus b times the quantity a minus b
$3! = 6$	three factorial equals 6

Differentials and Derivatives 微分と導関数

$f(x) = \sin x$	f of x equals sine x
$df = \cos x\, dx$	the differential of f equals cosine x d x
	d f equals cosine x d x
$\dfrac{df}{dx} = \cos x$	the derivative of f with respect to x equals cosine x
	d f by d x equals cosine x
$\dfrac{\partial u}{\partial x} = y\cos(xy)$	the partial derivative of u with respect to x equals y times cosine x y
	partial u by partial x equals y times cosine x y
$\dfrac{d^2 y}{dx^2} = -\sin x$	the second derivative of y with respect to x equals minus sine x
$\dfrac{\partial^2 u}{\partial x^2}$	the second partial derivative of u with respect to x
	second partial u by second partial x

$\dfrac{\partial^2 z}{\partial x \partial y}$	the second partial derivative of z with respect to x and y
$\dfrac{d^n y}{dx^n}$	the nth derivative of y with respect to x d n y by d x n
$\dfrac{d^2 y}{dx^2} = -\lambda^3 e^{-\lambda x}$	the second derivative of y with respect to x equals minus lambda cubed times e to the minus lambda times x power

Integrals 積分

$\displaystyle\sum_{i=1}^{n} a_i$	the sum from i equals one to n of a sub i
$\displaystyle\int k f(x) dx = k \int f(x)$	the indefinite integral of k times f of x with respect to x equals k times the indefinite integral of f of x
$\displaystyle\int k x^2 dx = \dfrac{k x^3}{3} + C$	the indefinite integral of k times x squared with respect to x equals the quantity k times x cubed over 3 plus C
$\displaystyle\int_a^b f(x) dx$	the integral from a to b of f of x
$\displaystyle\iint f(x,y) dx dy$	the double integral of f of x, y
$\displaystyle\int_c^d \int_a^b f(x,y) dx dy$	the integral from c to d with respect to y of the integral from a to b of f of x,y with respect to x
$\displaystyle\int \dfrac{2}{x+1} dx$ $= 2\log|x+1|$	the indefinite integral of the quantity two over x plus one with respect to x equals two times the quantity log of the absolute value of x plus one

$$\int_0^{\frac{\pi}{2}} (1 + k \cos x) dx$$

the integral from zero to pi over two of the quantity one plus k times cosine x with respect to x equals pi over two plus k

$$= \frac{\pi}{2} + k$$

Symbols (記号)

$x_k \to 0$ as $k \to \infty$	x sub k goes to 0 as k goes to infinity
$F(x) \equiv 0$	the function F of x is identical to 0
$F(x) \not\equiv G(x)$	the function F of x is not identical to G of x
$\hat{x} \approx x$	x hat is approximately equal to x
$\bar{x} \neq x^*$	x bar is not equal to x star (or x asterisk)
$x > y$	x is greater than y
$\tilde{x} < x'$	x tilde is less than x prime
$x < y$	x is less than y
$x \geq y$	x is greater than or equal to y
$x \leq y$	x is less than or equal to y
$x \in D$	x is an element of the set D
$D \subseteq X$	D is a subset of X
$\{x_k\} \subset D$	the sequence x sub k belongs to the set D
$\{x_{k_j}\} \subset \{x_k\}$	x sub k sub j is a subsequence of the sequence x sub k
$D = X \cap Y$	D is the intersection of X and Y
$D = X \cup Y$	D is the union of X and Y
$X \cap Y = \emptyset$	the intersection of X and Y is empty
$X \cap Y \neq \emptyset$	the intersection of X and Y is not empty
$\lim_{k \to \infty} \frac{1}{k}(a^k + b^{2/3})$	the limit as k goes to infinity of the quantity one over k times the quantity a to the kth power plus b to the two-thirds power
$f(x + h) = O(h)$	f of x plus h is of order h as h goes to 0

付録B　ギリシャ文字

Greek Alphabet

A	α	alpha	（アルファ）	N	ν	nu	（ニュー）
B	β	beta	（ベータ，ビータ）	Ξ	ξ	xi	（クサイ，クシー）
Γ	γ	gamma	（ガンマ）	O	o	omicron	（オミクロン）
Δ	δ	delta	（デルタ）	Π	π	pi	（パイ）
E	ϵ	epsilon	（イプシロン）	P	ρ	rho	（ロー）
Z	ζ	zeta	（ジータ）	Σ	σ	sigma	（シグマ）
H	η	eta	（イータ）	T	τ	tau	（タウ）
Θ	θ	theta	（シータ）	Υ	υ	upsilon	（ウプシロン）
I	ι	iota	（イオタ）	Φ	ϕ	phi	（ファイ）
K	κ	kappa	（カッパ）	X	χ	chi	（キー，カイ）
Λ	λ	lambda	（ラムダ）	Ψ	ψ	psi	（プサイ）
M	μ	mu	（ミュー）	Ω	ω	omega	（オメガ）

参 考 文 献

[1] G. Alefeld and J. Herzberger：Introduction to Interval Computations, Academic Press, Inc., New York (1983)
[2] K.E. Atkinson：An Introduction to Numerical Analysis, 2nd Edition, John Wiley & Sons, Inc., Toronto (1989)
[3] R.L. Burden and J.D. Farires：Numerical Analysis, 5th Edition, PWS-Kent Pub. Co., Boston (1993)
[4] S.J. Farlow：Partial Differential Equations for Scientists & Engineers, John Wiley & Sons, Inc. (1982)(伊理正夫・伊理由美訳：偏微分方程式，啓学社 (1984))
[5] F.B. Hildebrand：Introduction to Numerical Analysis, 2nd Edition, Dover Publications, Inc., New York (1974)
[6] E. Isaacson and H.B. Keller：Analysis of Numerical Methods, John Wiley & Sons, Inc., New York (1966)
[7] J.H. Mathews and K.D. Fink：Numerical Methods Using Matlab, 3rd Edition, Prentice-Hall, Inc., Upper Saddle River (1999)
[8] R. Kress：Numerical Analysis, Springer-Verlag, New York (1998)
[9] E. Kreyszig：Advanced Engineering Mathematics, 7th Edition, John Wiley & Sons, Inc., New York (1993)
[10] J.M. Ortega and W.C. Rheinboldt：Iterative Solution of Nonlinear Equations in Several Variables, Academic Press, Inc., New York (1970)
[11] L.B. Rall：Computational Solution of Nonlinear Operator Equations, John Wiley & Sons, Inc., New York (1969)
[12] 久保田光一，伊理正夫：アルゴリズムの自動微分と応用，コロナ社 (1998)
[13] 中尾充宏，山本野人：精度保証付き数値計算，日本評論社 (1998)
[14] 大石進一：数値計算，裳華房 (1999)
[15] 大石進一：精度保証付き数値計算，コロナ社 (2000)
[16] 山本哲朗：数値解析入門 (増訂版)，サイエンス社 (2003)
[17] 山本哲朗，北川高嗣：数値解析演習，サイエンス社 (1991)

演習問題略解

1 章

(1.1) $|x-a| \leq \varepsilon$ と $|x-a|/|x| \leq \varepsilon_R$ のような実数 ε と ε_R を求める。例えば、(1.1.1) において、$|\pi - 3.1415926| \leq 10^{-7}$, $|\pi - 3.1415926|/|\pi| \leq 10^{-7}/3.14$ より、$a = 3.1415926$ に対して、絶対誤差と相対誤差の誤差限界はそれぞれ 10^{-7} と $10^{-7}/3.14$.

(1.2) $|x-a| \leq 10^{-2}$ を満たす5けた（小数点以下4けた）の小数 a を計算すればよい。例えば、$|\pi - 3.1415926| \leq 10^{-2}$ より、$a = 3.1415$.

(1.3) $\dfrac{\hat{y}}{\hat{y}+e_2}$ の e_2 についてのテイラー展開式によって

$$\frac{\hat{y}}{\hat{y}+e_2} \approx 1 - \frac{e_2}{\hat{y}}.$$

したがって

$$\begin{aligned} \frac{x}{y} &= \frac{\hat{x}+e_1}{\hat{y}+e_2} \\ &\approx \frac{\hat{x}+e_1}{\hat{y}}\left(1 - \frac{e_2}{\hat{y}}\right) \\ &\approx \frac{\hat{x}}{\hat{y}} + \frac{e_1}{\hat{y}} - \frac{\hat{x}e_2}{\hat{y}^2} \end{aligned}$$

ゆえに

$$\left(\frac{x}{y} - \frac{\hat{x}}{\hat{y}}\right)\frac{y}{x} \approx \left(\frac{e_1}{\hat{y}} - \frac{\hat{x}e_2}{\hat{y}^2}\right)\frac{y}{x} \approx \frac{e_1}{x} - \frac{e_2}{y}.$$

(1.5) $k \geq 2$ のとき

$$\frac{a_{n+k}}{a_{n+1}} = \frac{a_{n+2}}{a_{n+1}} \cdot \frac{a_{n+3}}{a_{n+2}} \cdots \frac{a_{n+k}}{a_{n+k-1}} \leq \left(\frac{a_{n+2}}{a_{n+1}}\right)^{k-1}$$

であるから

$$\alpha = \frac{a_{n+2}}{a_{n+1}}$$

とおくとき

$$\begin{aligned} S - S_n &= a_{n+1} + a_{n+2} + a_{n+3} + \dots \\ &= a_{n+1}\left(1 + \frac{a_{n+2}}{a_{n+1}} + \frac{a_{n+3}}{a_{n+1}} + \dots\right) \\ &\leq a_{n+1}(1 + \alpha + \alpha^2 + \dots) \\ &= \frac{a_{n+1}}{1 - \alpha} \\ &= \frac{a_{n+1}^2}{a_{n+1} - a_{n+2}} \end{aligned}$$

(1.6) $f(x+h)$ と $f(x-h)$ を x を中心として 4 次の項までテイラー展開せよ。

2 章

(2.1) 解 $x^* = [1, 1, 1]$.
$$L = \begin{bmatrix} 1 & 0 & 0 \\ -\frac{1}{2} & 1 & 0 \\ 0 & -\frac{2}{3} & 1 \end{bmatrix}, \quad U = \begin{bmatrix} 2 & -1 & 0 \\ 0 & \frac{3}{2} & -1 \\ 0 & 0 & \frac{4}{3} \end{bmatrix}$$

(2.2) $x^* \approx [0.2500, -0.4375, -0.5625]$.

(2.3) $x^* = [1, 1, 0]$.
$$L = \begin{bmatrix} 2 & 0 & 0 \\ -1 & 1 & 0 \\ 0 & -1 & 1 \end{bmatrix}.$$

(2.8) Jacobi 法と Gauss-Seidel 法は，任意の初期値から (2.1) における連立 1 次方程式の一意解に収束する。

(2.9) $\|x\|_2$ について，$\sum_{i=1}^{n} x_i y_i \leq \|x\|_2 \|y\|_2$ に注意。

(2.11) $\|x\|_\infty = \max_{1 \leq i \leq n} |x_i| \neq 0$ の場合を考えればよい ($x = 0$ のときは明らか)。このとき

$$\|x\|_\infty \leq \|x\|_p = \|x\|_\infty \left(\sum_{i=1}^{n} \left(\frac{|x_i|}{\|x\|_\infty}\right)^p\right)^{1/p} \leq \|x\|_\infty n^{1/p}$$

において $p \to \infty$ とせよ。

(2.12),(2.13) Jacobi 法の反復行列のスペクトル半径 $\rho(D^{-1}(L+U))$ と Gauss-Seidel 法の反復行列のスペクトル半径 $\rho((D+L)^{-1}U)$ は 1 より小さいかどうかを考えよ。

(2.14) 係数行列が狭義優対角行列であることに注意。

3 章

(3.1) $g(x) = x - f(x)/f'(x)$ とする。任意の $x \in [1.2, 2.2]$ に対して，$|g'(x)| < 1$ を示せ。

(3.4) $f(x) = c - x^3$ とする。

(3.5) $f(x) = 3 - x^3$，$g(x) = x - f(x)/f'(x)$ とおく。任意の $x \in [a, b]$ に対して，$|g'(x)| < 1$ のような a, b を定める。

(3.6) $g(x) = x - f(x)/f'(x)$ とする。任意の $x \in [3/4a, 5/4a]$ に対して，$|g'(x)| < 1$ を示せ。

(3.7) 厳密解 $(x^*, y^*) = (1, 1)$．

(3.8) 厳密解 $(z_1^*, z_2^*, z_3^*) = (1, 2, 3)$．

4 章

(4.1) 対称行列のすべての固有値は実数であることに注意。

(4.2) $\lambda_i, i = 1, 2, \ldots, n$ を n 次行列 A の固有値とすれば，行列 $A - \mu I$ の固有値は $\lambda_i - \mu I, i = 1, 2, \ldots, n$．

(4.4) Householder 行列 $P = I - 2uu^T$ を利用する。ただし，$u = (x-y)/\|x-y\|_2$．

(4.5)
$$Q = \begin{bmatrix} -1.0000 & 0 & 0 \\ 0 & -0.9487 & -0.3162 \\ 0 & -0.3162 & 0.9487 \end{bmatrix}$$

$$R = \begin{bmatrix} -4.0000 & 0 & 0 \\ 0 & -3.1623 & -1.5811 \\ 0 & 0 & 1.5811 \end{bmatrix}$$

5 章

(5.1) $p_2(x) = -x^2 + 7x - 4$．

(5.2) ニュートンの前進差分補間多項式
$$p_3(x) = 2x + x(x-1) - \frac{2}{3}x(x-1)(x-2)$$

ニュートンの後退差分補間多項式
$$p_3(x) = 8 + 2(x-3) - (x-3)(x-2) - \frac{2}{3}(x-3)(x-2)(x-1).$$

(5.3)
$$|f(x) - p_n(x)| \le \frac{1}{(n+1)!} f^{(n+1)!}(\xi) \prod_{i=0}^{n} |x - x_i|$$

を利用する。

演 習 問 題 略 解

(5.4) 関数 $f(x) \equiv 1$ の Lagrange 補間多項式を考えよ．
(5.5) $f(x)$ はたかだか n 次の多項式であるから，誤差項 $E_n(x) = f(x) - p_n(x)$ はたかだか n 次の多項式であり，$n+1$ 個の点 $x_0, x_1, \ldots, x_n$ において 0 となる．ゆえに $E_n(x) \equiv 0$．

6 章

(6.2) 中点公式，台形公式および Simpson 公式の誤差評価はそれぞれ式 (6.4), (6.5), (6.6) を利用する．

(6.3) 分点 $x_0 = 1, x_1 = 3/2, x_2 = 2$ に関する $f(x) = 1/x$ の Lagrange 補間多項式を $p_2(x)$ とすれば，$x \in [1,2]$ に対し

$$f(x) - p_2(x) = \frac{1}{3!} f^{(3)}(\xi)(x-x_0)(x-x_1)(x-x_2), \ 1 < \xi < 2$$

かつ $f^{(3)}(\xi) = -3!(1/\xi^4)$．ゆえに $|f^{(3)}(\xi)| \leq 3!$ であって

$$I\left(\frac{1}{x}\right) - I_2\left(\frac{1}{x}\right) = \int_1^2 \{f(x) - p_2(x)\} dx,$$

$$\left| I\left(\frac{1}{x}\right) - I_2\left(\frac{1}{x}\right) \right| \leq \int_1^2 |f(x) - p_2(x)| dx \leq \int_1^2 \left|(x-1)\left(x - \frac{3}{2}\right)(x-2)\right| dx.$$

7 章

(7.1) 厳密解 $y(x) = e^x + x - 1$
(7.3) 厳密解 $y(x) = e^x + 1$
(7.4) 厳密解 $y(x) = e^x + \sin x$．

8 章

(8.1) 厳密解 $y(x) = e^x + x + 1$
(8.2) 厳密解 $u(x,y) = x^2 y^2$．
(8.3) 厳密解 $u(x,y) = 2e^{\pi x} \sin \pi y + (xy)^2$．
(8.4) 厳密解 $u(x,y) = \sin \pi x + \cos \pi y$．
(8.6) 厳密解 $u(x,t) = e^{(-\pi/2)^2 t} \sin \frac{\pi}{2} x$
(8.7) 厳密解 $u(x,t) = \sin \pi x \cos \pi t$

英和索引

A

absolute error ……………………1
　絶対誤差
accuracy ……………………………6
　精　度
Adams-Bashforth method ………109
　アダムス・バッシュフォース法
algebraic equation ………………36
　代数方程式
approximate value ………………1
　近似値
augumented matrix ………………15
　拡大行列
automatic differentiation ………43
　自動微分

B

back substitution …………………12
　後退代入
backward difference ……………77
　後退差分
boundary value problem …………115
　境界値問題

C

centered difference ………………116
　中心差分
characteristic determinant ………53
　特性行列式
characteristic equation ……………54
　特性方程式
characteristic polynomial …………54
　特性多項式
Cholesky's method ………………19
　コレスキー法

coefficient matrix ………………11
　係数行列
condition number ………………26
　条件数
contraction mapping ……………41
　縮小写像（縮小作用素）
contraction number ……………41
　縮小数
convergence ……………………23, 30
　収　束

D

derivative …………………………43
　微　分
determinant ………………………54
　行列式
diagonal matrix …………………20
　対角行列
diagonally dominant ………………32
　優対角（対角優位）
difference table …………………74
　差分表
direct method ……………………11, 132
　直接法
Dirichlet problem ………………121
　ディリクレ問題
discretization error ………………107
　離散化誤差
divided differences ………………73
　差分商
Durand-Kerner method …………50
　デュラン・カーナー法

E

eigenvalue…………………………53
　固有値

eigenvalue problem ········· 53
　　固有値問題
eigenvector ················· 53
　　固有ベクトル
elimination method ········· 11
　　消去法
elliptic equation ············ 121
　　楕円型方程式
error ························· 1
　　誤　差
error bound ················· 2
　　誤差限界
error formula ················ 84
　　誤差公式
error propagation ············ 6
　　誤差の伝播
Euclidean norm ·············· 26
　　ユークリッドノルム
Euler method ················ 103
　　オイラー法
extrapolation ················ 67
　　外　挿

F

finite difference approximation ······ 119
　　有限差分近似
finite difference equation ············ 117
　　有限差分方程式
finite difference method ············· 115
　　有限差分法
five-point difference formula ········ 124
　　5点差分公式
fixed point ·················· 7, 37
　　不動点
fixed point method ··········· 37
　　不動点法
floating-point number ········ 7
　　浮動小数点数
forward differences ·········· 77
　　前進差分
forward elimination ·········· 12
　　前進消去

G

Gaussian elimination ········· 11
　　ガウスの消去法
Gauss-Seidel method ········· 21
　　ガウス・ザイデル法
Gerschgorin's theorem ······· 55
　　ゲルシュゴリンの定理
global truncation error ······ 104
　　大域打切り誤差

H

heat equation ················ 127
　　熱方程式
Householder matrix ········· 59
　　ハウスホルダー行列
hyperbolic equation ·········· 131
　　双曲型方程式

I

identity matrix ··············· 53
　　単位行列
ill-conditioned ················ 27
　　悪条件
ill posed problem ············· 102
　　悪条件問題，不安定問題
inclusion theorem ············ 56
　　包含定理
initial value problem ········· 101
　　初期値問題
instability ···················· 88
　　不安定性
interpolation ················· 67
　　補　間
interpolation polynomial ····· 67
　　補間多項式
inverse matrix ················ 16
　　逆行列
iterative method ············· 11
　　反復法

英和索引

J

Jacobian matrix ·······························45
 ヤコビ行列
Jacobi method ································21
 ヤコビ法

L

Lagrange factor ·······························69
 ラグランジュ因子
Lagrange interpolation polynomial ···69
 ラグランジュ補間多項式
limit ··23
 極　限
linear convergence ····························5
 1次収束（線形収束）
linear interpolating polynomial ········92
 線形補間多項式
Lipschitz condition ·······················102
 リプシッツ条件
l_1-norm ··24
 l_1-ノルム
l_2-norm ··24
 l_2-ノルム
l_∞-norm ·······································24
 l_∞-ノルム
local truncation error ·····················103
 局所打切り誤差
loss of significant digits ······················7
 けた落ち
lower triangular matrix ····················20
 下三角行列
LU-factorization ······························11
 LU分解

M

matrix norm ···································23
 行列ノルム
mean value theorem ·························88
 平均値定理
mesh point ···································116
 格子点（分点）
midpoint rule ·································89
 中点公式，中点則
multistep method ···························109
 多段法

N

Newton-Cotes quadrature formula ···88
 ニュートン・コーツ積分公式
Newton interpolation formula ······73, 77
 ニュートン補間公式
Newton method ·························41, 46
 ニュートン法
nonlinear equation ···························36
 非線形方程式
nonsingular matrix ···························26
 正則行列
norm ··23
 ノルム
numerical integration ·······················86
 数値積分

O

order of approximation ······················4
 近似の次数
order of convergence ························5
 収束次数（収束位数）
ordinary differential equation ·········101
 常微分方程式
orthogonal matrix ····························54
 直交行列

P

parabolic equation ·························127
 放物型方程式
partial differential equation ············122
 偏微分方程式
partial pivoting (strategy) ·················13
 部分ピボット（選択法）
pivot ··12
 ピボット
polynomial ······························36, 49
 多項式

positive definite matrix ·················· 19
　正定値行列
power method ····························· 57
　累乗法

Q

QR factorization ·························· 63
　QR 分解
QR method ································· 64
　QR 法
quadratic convergence ············· 5, 49
　2 次収束
quadrature formula ····················· 87
　数値積分公式，求積公式

R

rectangular rule ··························· 89
　矩形（長方形）公式
relative error ······························· 1
　相対誤差
root ·· 36
　根
rounding down ···························· 3
　切捨て
rounding up ································ 3
　切上げ
round-off error ···························· 2
　丸め誤差
Runge-Kutta method ················· 105
　ルンゲ・クッタ法

S

secant method ···························· 43
　割線法（セカント法）
sequence ··································· 20
　列
series ··· 4
　級数
significant digits ·························· 7
　有効けた
similar matrix ····························· 54
　相似行列

Simpson's rule ···························· 96
　シンプソンの公式
single-step method ···················· 109
　1 段法
spectral radius ··························· 28
　スペクトル半径
stability ··································· 129
　安定性
starting value ···························· 43
　出発値
step size ·································· 102
　刻み幅
strictly diagonally dominant ········· 32
　狭義優対角，狭義対角優位
superlinear convergence ··············· 5
　超 1 次（線形）収束
symmetric function ····················· 50
　対称関数
symmetric matrix ························ 54
　対称行列

T

Taylor expansion ·················· 4, 106
　テイラー展開
Taylor series ······························· 4
　テイラー級数
trace ··· 54
　トレース（跡）
trapezoidal rule ··························· 93
　台形公式
triangle inequality ······················· 23
　三角不等式
tridiagonal matrix ··················· 59, 64
　3 重対角行列
truncation error ··········· 4, 110, 118, 124
　打切り誤差

U

upper triangular matrix ················ 20
　上三角行列

V

Vandermonde matrix ·················68
　ファンデアモンド行列
vector norm ································23
　ベクトルノルム
verification method ·····················8
　精度保証法

W

wave equation ························134
　波動方程式

Weierstrass' method ················49
　ワイヤストラス法
weights ·····································86
　重　み

Z

zeros of polynomial ················49
　多項式の零点

和 英 索 引

あ 行

悪条件 ……………………………… 2
　ill-conditioned
悪条件問題 ……………………… 102
　ill posed ploblem
アダムス・バッシュフォース法 ……… 109
　Adams-Bashforth method
安定性 …………………………… 129
　stability
1次収束（線形収束） ……………… 5
　linear convergence
1段法 …………………………… 109
　single-step method
上三角行列 ……………………… 20
　upper triangular matrix
打切り誤差 ……………… 4, 110, 118, 124
　truncation error
l_1-ノルム ………………………… 24
　l_1-norm
l_2-ノルム ………………………… 24
　l_2-norm
l_∞-ノルム ………………………… 24
　l_∞-norm
LU 分解 …………………………… 11
　LU factorization
オイラー法 ……………………… 103
　Euler method

か 行

外挿 ……………………………… 67
　extrapolation
ガウス・ザイデル法 ……………… 21
　Gauss-Seidel method
ガウスの消去法 …………………… 11
　Gaussian elimination
拡大行列 ………………………… 15
　augumented matrix
割線法（セカント法） …………… 43
　secant method
刻み幅 …………………………… 102
　step size
基本対称関数 …………………… 50
　elementary symmetric function
逆行列 …………………………… 16
　inverse matrix
QR 分解 ………………………… 63
　QR factorization
級　数 …………………………… 4
　series
狭義優対角（狭義対角優位） ……… 32
　strictly diagonally dominant
行列式 …………………………… 54
　determinant
行列ノルム ……………………… 23
　matrix norm
極　限 …………………………… 23
　limit
局所打切り誤差 ………………… 104
　local trancation error
切上げ …………………………… 3
　rounding up
切捨て …………………………… 3
　rounding down
近似値 …………………………… 1
　approximate value
近似の次（位）数 ………………… 4
　order of appoximation
矩形公式 ………………………… 89
　rectangular rule
係数行列 ………………………… 11
　coefficient matrix

境界値問題 ……………………115
 boundary value problem
けた落ち ………………………7
 loss of significant digits
ゲルシュゴリンの定理 ……………56
 Gerschgorin's theorem
格子点（分点）……………………116
 mesh point
後退差分 ………………………77
 backward difference
後退代入 ………………………12
 back substitution
誤　差 …………………………1
 error
誤差の伝播 ………………………6
 error propagation
誤差の限界 ………………………2
 error bound
誤差公式 ………………………84
 error formula
5点差分公式 ……………………124
 five-point difference formula
固有値 …………………………53
 eigenvalue
固有値問題 ………………………53
 eigenvalue problem
固有ベクトル ……………………53
 eigenvector
コレスキー法 ……………………19
 Cholesky's method
根 ………………………………36
 root

さ　行

差分商 …………………………73
 divided difference
差分表 …………………………74
 difference table
有限差分方程式 …………………117
 finite difference equation
三角不等式 ………………………23
 triangle inequality

3重対角行列 ……………………59,64
 tridiagonal matrix
下三角行列 ………………………20
 lower triangular matrix
自動微分 ………………………43
 automatic differentiation
収　束 …………………………23,30
 convergence
収束次数（収束位数）………………5
 order of convergence
縮小写像（縮小作用素）……………41
 contraction mapping
縮小写像の原理 …………………41
 contraction mapping principle
縮小数 …………………………41
 contraction number
消去法 …………………………11
 elimination method
条件数 …………………………26
 condition number
出発値 …………………………43
 starting value
初期値問題 ……………………101
 initial value problem
常微分方程式 ……………………101
 ordinary differential equation
シンプソンの公式 ………………96
 Simpson's rule
数値積分 ………………………86
 numerical integration
数値積分公式 ……………………87
 quadrature formula
スペクトル半径 …………………28
 spectral radius
正則行列 ………………………26
 nonsingular matrix
正定値行列 ………………………19
 positive definite matrix
精　度 …………………………6
 accuracy
精度保証法 ………………………8
 verification method

和英索引

セカント法（割線法） ……………43
  secant method
絶対誤差 ……………………………1
  absolute error
線形収束 ……………………………5
  linear convergence
線形補間多項式 ……………………92
  linear interpolating polynomial
前進差分 ……………………………77
  forward difference
前進消去 ……………………………12
  forward elimination
双曲型方程式 ………………………131
  hyperbolic equation
相似行列 ……………………………54
  similar matrix
相対誤差 ……………………………1
  relative error

た 行

大域打切り誤差 ……………………104
  global truncation error
対角行列 ……………………………20
  diagonal matrix
台形公式 ……………………………93
  trapezoidal rule
対称関数 ……………………………50
  symmetric function
対称行列 ……………………………54
  symmetric matrix
代数方程式 …………………………36
  algebraic equation
楕円型方程式 ………………………121
  elliptic equation
多項式 ……………………………36,49
  polynomial
多項式の零点 ………………………49
  zeros of polynomial
多段法 ………………………………109
  multistep method
単位行列 ……………………………53
  identity matrix
中心差分近似公式 …………………116
  centered difference formula
中点公式 ……………………………89
  midpoint rule
超1次（線形）収束 ………………5
  superlinear convergence
直接法 …………………………11,128,132
  direct method
直交行列 ……………………………54
  orthogonal matrix
テイラー展開 ……………………4,106
  Taylor expansion
テイラー級数 ………………………4
  Taylor series
ディリクレ問題 ……………………121
  Dirichlet problem
デュラン・カーナー法 ……………50
  Durand-Kerner method
特性行列式 …………………………53
  characteristic determinant
特性多項式 …………………………54
  characteristic polynomial
特性方程式 …………………………54
  characteristic equation
トレース（跡） ……………………54
  trace

な 行

2次収束 …………………………5,49
  quadratic convergence
ニュートン・コーツ積分公式 ……88
  Newton-Cotes quadrature formula
ニュートン法 …………………41,46
  Newton's method
ニュートン補間公式 ……………73,77
  Newton interpolation formula
熱方程式 ……………………………127
  heat equation
ノルム ………………………………23
  norm

は 行

ハウスホルダー行列 …………………59
　Householder matrix
波動方程式 ………………………134
　wave equation
反復法 ………………………………11
　iterative method
微　分 ………………………………43
　derivative
非線形方程式 ………………………36
　nonlinear equation
ピボット ……………………………12
　pivot
ファンデアモンド行列 ……………68
　Vandermonde matrix
不安定性 ……………………………88
　instability
浮動小数点数 ………………………7
　floating-point number
不動点 ………………………………37
　fixed point
不動点法 ……………………………37
　fixed point method
部分ピボット ………………………13
　partial pivoting
平均値定理 …………………………88
　mean value theorem
ベクトルノルム ……………………23
　vector norm
偏微分方程式 ………………………122
　partial differential equation
包含定理 ……………………………56
　inclusion theorem
放物型方程式 ………………………127
　parabolic equation
補　間 ………………………………67
　interpolation
補間多項式 …………………………67
　interpolation polynomial

ま 行

丸め誤差 ……………………………3
　round-off error

や 行

ヤコビ法 ……………………………21
　Jacobi method
ヤコビ行列 …………………………45
　Jacobian matrix
ユークリッドノルム ………………26
　Euclidean norm
有限差分近似 ………………………119
　finite difference approximation
有限差分法 …………………………115
　finite difference method
有効けた ……………………………7
　significant digits
優対角 ………………………………32
　diagonally dominant

ら 行

ラグランジュ因子 …………………69
　Lagrange factor
ラグランジュ補間多項式 …………69
　Lagrange interpolation polynomial
リプシッツ定数 ……………………102
　Lipschitz constant
リプシッツ条件 ……………………102
　Lipschitz condition
累乗法 ………………………………57
　power method
ルンゲ・クッタ法 …………………105
　Runge-Kutta method
列 ……………………………………20
　sequence

わ 行

ワイヤストラス法 …………………49
　Weierstrass' method

―― 著者略歴 ――

陳　小君（ちん　しょうくん）
1987 年	西安交通大学大学院博士課程修了（数学科）
	理学博士（西安交通大学）
1990 年	理学博士（岡山理科大学）
1991 年	New South Wales 大学研究員
1998 年	島根大学助教授
2002 年	弘前大学教授
2008 年	香港理工大学教授
	現在に至る

山本哲朗（やまもと　てつろう）
1961 年	広島大学大学院修士課程修了（数学専攻）
1966 年	広島大学講師
1968 年	理学博士（広島大学）
1969 年	愛媛大学助教授
1975 年	愛媛大学教授
2002 年	愛媛大学名誉教授
2002 年	早稲田大学客員教授
2007 年	早稲田大学退職

英語で学ぶ**数値解析**
Numerical Analysis　　　　　　　　　　　© Xiaojun Chen, Tetsuro Yamamoto 2002

2002 年 10 月 18 日　初版第 1 刷発行
2017 年 3 月 10 日　初版第 4 刷発行

検印省略

著　者	陳　　　小　　　君
	山　本　哲　朗
発行者	株式会社　コロナ社
	代表者　牛来真也
印刷所	壮光舎印刷株式会社
製本所	株式会社　グリーン

112-0011　東京都文京区千石 4-46-10
発行所　株式会社　コロナ社
CORONA PUBLISHING CO., LTD.
Tokyo Japan
振替00140-8-14844・電話(03)3941-3131(代)
ホームページ　http://www.coronasha.co.jp

ISBN 978-4-339-06072-0　C3041　Printed in Japan　　　　　　（金）

〈出版者著作権管理機構　委託出版物〉
本書の無断複製は著作権法上での例外を除き禁じられています。複製される場合は，そのつど事前に，出版者著作権管理機構（電話 03-3513-6969，FAX 03-3513-6979，e-mail: info@jcopy.or.jp）の許諾を得てください。

本書のコピー，スキャン，デジタル化等の無断複製・転載は著作権法上での例外を除き禁じられています。購入者以外の第三者による本書の電子データ化及び電子書籍化は，いかなる場合も認めていません。
落丁・乱丁はお取替えいたします。

コンピュータ数学シリーズ

(各巻A5判，欠番は品切です)

■編集委員　斎藤信男・有澤　誠・筧　捷彦

配本順			頁	本体
2.(9回)	組合せ数学	仙波一郎著	212	2800円
3.(3回)	数理論理学	林　晋著	190	2400円
7.(10回)	ゲーム計算メカニズム ―将棋・囲碁・オセロ・チェスのプログラムはどう動く―	小谷善行編著	204	2800円
10.(2回)	コンパイラの理論	大山口通夫著	176	2200円
11.(1回)	アルゴリズムとその解析	有澤　誠著	138	1650円
16.(6回)	人工知能の理論（増補）	白井良明著	182	2100円
20.(4回)	超並列処理コンパイラ	村岡洋一著	190	2300円
21.(7回)	ニューラルコンピューティング	武藤佳恭著	132	1700円

以下続刊

1.	離散数学	難波完爾著	4.	計算の理論　町田元著
5.	符号化の理論	今井秀樹著	6.	情報構造の数理　中森真理雄著
8.	プログラムの理論		9.	プログラムの意味論　荻野達也著
12.	データベースの理論		13.	オペレーティングシステムの理論　斎藤信男著
14.	システム性能解析の理論	亀田壽夫著	17.	コンピュータグラフィックスの理論　金井崇著
18.	数式処理の数学	渡辺隼郎著	19.	文字処理の理論

定価は本体価格＋税です。
定価は変更されることがありますのでご了承下さい。

図書目録進呈◆